Finite Element Thermography

Finite Element Thermography

Benoit Beckers
Pierre Beckers

iSTE

WILEY

First published 2026 in Great Britain and the United States by ISTE Ltd and John Wiley & Sons, Inc.

ISTE Ltd
27-37 St George's Road
London SW19 4EU
UK

www.iste.co.uk

John Wiley & Sons, Inc.
111 River Street
Hoboken, NJ 07030
USA

www.wiley.com

Library of Congress Control Number: 2026935942

British Library Cataloguing-in-Publication Data
A CIP record for this book is available from the British Library
ISBN 978-1-83669-078-8

The manufacturer's authorized representative according to the EU General Product Safety Regulation is Wiley-VCH GmbH, Boschstr. 12, 69469 Weinheim, Germany, e-mail: Product_Safety@wiley.com.

Contents

Preface

This book is primarily intended for young researchers who use thermal cameras in laboratories or outdoors to observe the behavior of machines, industrial processes or civil engineering structures, and who, faced with the considerable difficulties of interpreting thermographic images, wish to simulate such images on a computer. They are led naturally to the finite element method, but they discover that existing codes are ill-suited to this type of problem and that it is necessary to develop procedures for pre-processing, calculation and post-processing for which there are no guides or truly relevant reference works.

In this preface, however, we would first like to address experienced colleagues, mainly from the finite element community, who are likely to supervise such work or teach the theory behind it, and who may feel somewhat disconcerted by the variety of scientific sources that need to be consulted in order to properly address a problem that appears so simple. It is indeed simple, as it essentially involves coupling two well-known equations: the equation of heat diffusion in solids, a partial differential equation that is elliptic in steady state and parabolic in transient state, and, on the other hand, for the reflections of radiation between the surfaces of the scene, the integral equation that bears the name of one of the two mathematicians who described it simultaneously at the beginning of the 20th century, Ivar Fredholm (1866–1947) and Vito Volterra (1860–1940). The latter lends itself very well to the Monte Carlo method. A staggering number of rays must be launched from all surfaces of the scene to obtain an accurate solution, but this is no longer really a problem today, thanks to advances in computing.

The researchers can therefore be imagined constructing the geometry to be simulated and meshing it using an automatic process. This is probably done in an unstructured manner, using tetrahedrons from which the skin must then be extracted. The skin itself consists of a large number of triangles, which may be subject to reflections from shortwave radiation, for example, from the Sun (first ray tracing, independent of the finite element calculation). Long-wave radiation is then emitted according to the fourth power of their temperature, which in turn is reflected (second ray tracing, this time integrated into the finite difference scheme). Once the calculation is complete, a third ray tracing, performed in post-processing from the camera position, finally allows the desired image to be formed. In addition to numerous technical difficulties, however, this approach leads to theoretical dissatisfaction.

The idealization of nature that serves as the framework for images (visible world), Fourier's equation (thermal gradients) and the finite element method is that of *continuous media*. It corresponds to the scale at which human beings live and develop their artifacts, from the smallest, which fit in the palm of the hand, to the largest works of architecture and civil engineering. In order to faithfully follow the shape of an object, the finite element has a size that typically varies from a tenth of a millimeter to a hundred meters. On either side of this scale, geometry, in the Greek sense, is lost in statistical and then discrete descriptions. Here, we bring back from quantum mechanics the idealization of the continuity of the black body, while, from Keplerian orbits, we report the apparent paths of the Sun in the continuous sky of our scene.

In continuous media, the variational calculus allows us to solve a partial differential equation globally based on a functional that is generally minimized. Compared to Galerkin's finite element formulation, which discretizes not only the geometry but also the equation, the variational calculus only discretizes the functional. This is why we systematically use the latter, successively introducing into the functional of the temperature field steady-state diffusion, Neumann boundary conditions and convection, but also transient behavior and radiation, which is less common in the literature. Compared to the clarity of this presentation, the Monte Carlo solution of the Volterra equation seems all the more obscure: each ray emitted ignores its neighbors, and we are no longer in continuous media.

In the second half of the 20th century, the emerging community of computer graphics and realistic rendering took up this equation, renaming it the "global illumination equation". Two main assumptions – Lambertian reflection and the scene being discretized into patches – allow it to be reduced to a simple linear system, where all the geometry is concentrated in the "view factors". For the rest, the *radiosity method* is so similar to the finite element method that some authors do not hesitate to relate it to the latter. In fact, it has been included quite naturally in the main finite element codes, but in its classical form, with the restrictive assumption of perfectly diffuse reflection.

However, when the scenes studied become complex, it is necessary to reintroduce ray tracing techniques, but only in the evaluation of view factors, in order to accurately detect objects that lie between two patches (elimination of hidden parts). There is then nothing to prevent a ray that encounters a polished surface from being extended by specular reflection, so as to extend the view factor to "the other side of the mirror". In fact, *extended view factors* make it possible to take into account all possible types of reflection or transmission through translucent objects. It is remarkable that discretization now only operates on geometry: discretization of the scene's surfaces into patches, which form the skin of the finite element mesh, then discretization of view factors into rays, without any loss of generality in the description of optical properties. There is therefore a great affinity between the variational calculus and the extended radiosity method: the equations and the physics they describe remain intact in the idealization of continuous media.

In the last third of the 20th century, another simplification came, this time from the computer-aided design community. As we recall here, this idea took a few years to be taken up by the emerging finite element community under the name of *isoparametric elements*: the same shape functions describe both the geometry and the physical field under study. In this book, we show the advantages of the resulting structured mesh, particularly for conductive-radiative coupling. Based on our previous work, we propose a partition of the sphere or hemisphere that allows ray tracing to be stratified in an ideal manner, either at equal solid angles or at equal view factors. When drawing from the skin of isoparametric elements, we still benefit from the properties of structured, and therefore indexed, meshing for element tracking. Structured meshing also allows us to perfectly respect the symmetries of the geometry under study. If the boundary conditions imposed are also symmetric, then these symmetries can be found in the solution, which is a key advantage for understanding, interpreting and verifying simulations.

In Chapter 8, we borrow from a very specific community involved in comfort in buildings, the concept of *mean radiant temperature*, which is the temperature that would be measured by a thermometer placed in a black globe. This temperature is defined at every point in space, even in a vacuum or a perfectly transparent medium. It therefore illustrates well the integration of radiation into continuous media and should find its place in radiation physics, beyond the limited field in which it was proposed.

This book therefore stops short of introducing fluid dynamics and other complex phenomena, such as phase changes. However, it should be noted that such problems, while certainly very difficult to implement in simulation, do not pose any particular difficulty in terms of their participation in continuous media. This difficulty only existed for radiation. We have clarified this here, and we believe that our approach is the best possible starting point for any future multiphysics application based on the thermodynamics of continuous media.

One final comment concerns multiscale simulations, which have become commonplace in other fields of mechanics, such as elasticity, but are still poorly defined in the field of thermal engineering. The scale we describe here is centered on a mesh with sides measuring 10 cm to within a factor of 10. At the lower scale, centered on the millimeter, it becomes inconceivable to mesh an entire scene. This is the first scale of materials, such as the thickness of a window pane that we want to give certain thermal and optical properties. At our scale, it is common to receive such properties in order to parameterize simulations, for instance, of a large building using this glass. We should return the results of our simulations as boundary conditions for material-specific studies, but this has not yet been done. There is therefore no dialogue with the lower scale, nor with the scale immediately above it. The latter, centered on the decameter, is still part of the continuous environment, but the geometry is necessarily simplified. It is probably indispensable for certain fluid dynamics simulations that are still unfeasible on large models (e.g. the entire city where the building is located), which would be too detailed. This scale should in turn interact with the two higher scales, clearly identified as the *meso* scale (1 km grid, used by meteorologists, among others) and the *macro* scale (100 km grid, used by climatologists for global simulations of our planet).

Continuous environments are vast and their description involves three spatial scales separated from each other by a factor of 100: millimeters, decimeters and decameters. If we start from the central scale – for example, to study heat exchange in a street – we need to consider the two scales on either side. In our example, these are the scale of coatings and the scale of the entire city, both of which still belong to continuous environments. Everything explained in this book therefore still applies. The same program can be used for all three scales, provided that it is possible to zoom in and out of the geometric model, which can be achieved using adaptive meshing techniques. On the other hand, until such a process is mastered, it would be premature to force dialogue with meteorologists and materials physicists: we have only begun to clear the way, and with this book we are laying a first, but essential, milestone.

March 2026

Introduction

In the physical world, heat is transmitted in two very different ways: from one point to another within what are known as continuous media, and by radiation in transparent media. In the first mode, particles move within a solid (conduction) or within a fluid (convection). In the second, heat is transported by photons across immense distances through a vacuum at the speed of light, eventually entering the Earth's atmosphere and other semi-transparent media, such as water or glass, with varying degrees of interaction in the form of absorption and diffusion (blue sky, clouds, the ocean and windows).

This book is divided into two parts. The first focuses on the heat diffusion equation and the second on the radiosity equation. Finite element methods are the preferred approach for solving the former, while ray tracing is used for the latter. These two methods have rarely been addressed together. The main focus of this book is to present the calculation of the coupling between conduction and radiation for complex geometries, with all the necessary generality in terms of describing the thermal properties of materials, as well as the optical properties of the surfaces that enclose them.

Chapter 1 describes the finite element method, following the five historical steps that led to its development over two centuries. First, we recall the partial differential equation of diffusion proposed by Joseph Fourier in the early 19th century. Next, we look at the steady-state functional, which must be minimized using the techniques of variational calculus developed by Leonhard Euler and Joseph-Louis Lagrange at the end of the previous century. It was not until a century later, in 1908, that Walther

Ritz proposed the test functions that enabled substantial progress in solving the problem. In the mid-20th century, the development of computers made it possible to divide the geometry under study into a large number of finite-sized elements and apply a procedure known as Rayleigh–Ritz to each of them. In the mid-1960s, Bruce Irons, building on ideas developed by Steven Anson Coons in the field of computer-aided design, defined isoparametric elements, which give the finite element method its most elegant formulation in the context of structured meshing.

In Chapter 2, boundary conditions are addressed using Lagrange multipliers, which provide a very practical form for Dirichlet conditions and also allow non-coincident structured meshes to be joined together. Neumann conditions are then discussed, as well as convection, based on virtual nodes related to Coons patches, which are the surfaces limiting the isoparametric elements that make up the solid under study. In these different aspects, we take advantage of the particular properties of structured meshes, which will accompany us until the end of the book. The same problems can be studied with the ultimate unstructured mesh, which is composed of Delaunay tetrahedra and triangles. However, the differences are so numerous that a second book would be required to arrive at the same results, albeit with considerable difficulty, notably in expressing the coupling between conduction and radiation. The common thread of our reasoning is provided by isoparametric elements, whose appeal we wish to restore, particularly in the field of thermal engineering.

Chapter 3 proposes an extension of the variational principle to transient regimes, as well as a finite difference scheme to address the temporal aspect. Thermal capacity is introduced, and we see that thermal inertia causes considerable delays in the diffusion of heat through solids. It is therefore conduction that generally governs thermal problems.

Isoparametric elements are detailed in Chapter 4. Their name derives from the fact that the shape functions are the same for describing the geometry and for describing the physical field being studied. They allow complex geometries to be reproduced, preserving their symmetries, with easy identification of nodes and elements.

The second part of the book begins in Chapter 5 with a description of the spectral density of electromagnetic radiation emitted by an ideal black body

(Planck's law), showing the separation between short waves (emitted by the Sun) and long waves (emitted by the Earth). The emphasis is on the spatial description of radiation, which draws on two fundamental geometric concepts: solid angle and viewing factor. Finally, we show how to efficiently calculate the viewing factor between two Coons patches.

Short waves are discussed in Chapter 6. We briefly review the apparent paths of the Sun in the sky and the attenuation of its radiation by the atmosphere. The reflections of short waves on the surfaces of a terrestrial scene are studied in general terms, whether the reflection is perfectly diffuse (classical formulation of radiance equations) or more complex (introduction of extended view factors to take into account mirrors, for example). Since shortwave radiation is perfectly decoupled from that of objects in the scene (which only emit long waves), the resulting irradiance on Coons patches is reduced to a simple load.

The same cannot be said for long waves, which are discussed in Chapter 7. In this case, all surfaces in the scene emit radiation based on their temperature (Stefan–Boltzmann law), which is calculated using finite elements. In transient conditions, this calculation must therefore be integrated into the finite difference scheme. At the end of this chapter, it has become possible to obtain perspective views of the scene, where each surface is colored according to its temperature, in other words, a digital version of thermography.

In the eighth and final chapter, we propose to create a panoramic thermogram of the scene using double projection. The scene is first projected onto a unit sphere surrounding the viewpoint. This central projection is then carried to the plane, following the technique proposed in 1805 by Karl Brandan Mollweide. The Mollweide projection is equivalent, meaning that it preserves surface ratios and therefore solid angles, which are, by definition, the areas projected onto the unit sphere and measured in steradians. This means that the radiances emanating from the surfaces of the scene toward the viewpoint can be directly added to the pixels of the Mollweide projection to obtain the mean radiant temperature. This, averaged with the air temperature, gives the operative temperature, which ultimately provides a good indication of how a person standing in that location would feel in the absence of wind.

This book offers a study of the coupling between conduction and radiation, which is relevant to many applications. The separation between short waves and long waves is not always relevant, as it only applies to scenes located on our planet and subject to solar radiation (Beckers 2016). Among these scenes, the most stimulating are cities and the buildings that form them, which, due to their regular shapes, find their natural mesh in isoparametric elements. This book is therefore particularly intended for simulations of urban physics.

Part 1

Thermal Diffusion

1

Steady State

Thermal diffusion is described by a partial differential equation, expressed here in matrix notation, which implies certain conventions. Vectors are systematically represented by column matrices. In general, developments are carried out in three dimensions. These are the same as those detailed in the report accompanying the *Fiammetta* program (Beckers 2023), which illustrates all the concepts in the first part of this book, but in two dimensions. In this report, interested readers can follow a step-by-step description of the code in English and extract all the Fiammetta routines written in Matlab®.

1.1. Partial differential equations of heat transfer in a solid

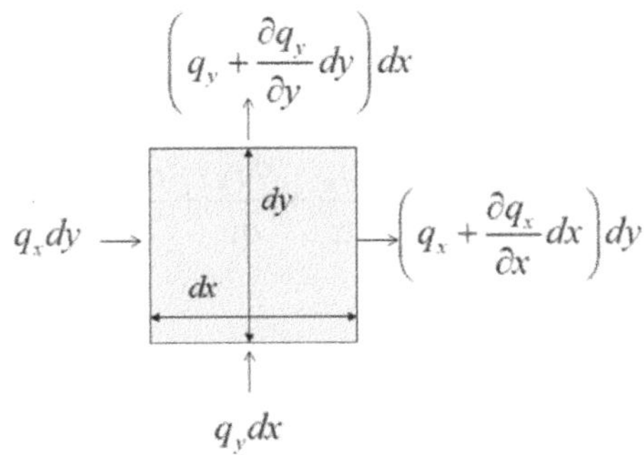

Figure 1.1. *Two-dimensional local heat flow balance in steady state*

For a color version of all the figures in this chapter, see www.iste.co.uk/beckers/finite.zip.

The basic physical entity is *thermal flux* or heat flux. Its components are the fluxes directed along the x-axis: q_x, the y-axis: q_y, and the z-axis: q_z.

The first step is to show the link between the partial differential equation satisfied by the flows and the local equilibrium of heat flows in steady state.

To facilitate graphical representation, Figure 1.1 shows the thermal flow balance in two dimensions. In the absence of heat sources or sinks within the differential element, the sum of the four incoming and outgoing flows of the differential element *dx dy* is equal to zero:

$$\left(q_y + \frac{\partial q_y}{\partial y} dy\right) dx + \left(q_x + \frac{\partial q_x}{\partial x} dx\right) dy - q_x dy - q_y dx = 0$$

or $$\left(\frac{\partial q_y}{\partial y} + \frac{\partial q_x}{\partial x}\right) dxdy = 0 \qquad [1.1]$$

This is the *continuity equation* or two-dimensional equilibrium equation: the divergence of the heat flux is equal to zero. The nabla operator is defined as the column matrix of partial derivatives:

$$\vec{\nabla}^T = \left[\frac{\partial}{\partial x}, \frac{\partial}{\partial y}, \frac{\partial}{\partial z}\right] \qquad [1.2]$$

The scalar product of the nabla operator and the column matrix of heat fluxes gives a scalar called the *divergence of the heat flux*:

$$\vec{\nabla}^T . \vec{q} = \begin{bmatrix} \frac{\partial}{\partial x} & \frac{\partial}{\partial y} & \frac{\partial}{\partial z} \end{bmatrix} \begin{bmatrix} q_x \\ q_y \\ q_z \end{bmatrix} = \frac{\partial q_x}{\partial x} + \frac{\partial q_y}{\partial y} + \frac{\partial q_z}{\partial z} \qquad [1.3]$$

The divergence is the product of a row matrix by a column matrix, the result being a scalar [1.3]. In the presence of heat source distribution $\vec{Q}$ imposed in the volume, the equilibrium equation is written as:

$$\vec{\nabla}^T . \vec{q} + \vec{Q} = 0 \qquad [1.4]$$

The product of the nabla and a scalar, for example, temperature, is called the *gradient*; it results in the column matrix of the thermal gradient:

$$\vec{\nabla}\tau = \begin{bmatrix} \frac{\partial}{\partial x} \\ \frac{\partial}{\partial y} \\ \frac{\partial}{\partial z} \end{bmatrix} \tau = \begin{bmatrix} \frac{\partial \tau}{\partial x} \\ \frac{\partial \tau}{\partial y} \\ \frac{\partial \tau}{\partial z} \end{bmatrix} \qquad [1.5]$$

The Fourier equation links the temperature field τ expressed in degrees Kelvin (K) to the vector $\vec{q}$ (column matrix) of heat flux expressed in W m^{-2} (Fourier 1822).

In an isotropic medium, the heat flux is proportional to the temperature gradient and the thermal conductivity k (W m^{-1} K^{-1}):

$$\vec{q} = -k\,\vec{\nabla}\tau \rightarrow \begin{bmatrix} q_x \\ q_y \\ q_z \end{bmatrix} = -k \begin{bmatrix} \frac{\partial}{\partial x} \\ \frac{\partial}{\partial y} \\ \frac{\partial}{\partial z} \end{bmatrix} \tau = -k \begin{bmatrix} \frac{\partial \tau}{\partial x} \\ \frac{\partial \tau}{\partial y} \\ \frac{\partial \tau}{\partial z} \end{bmatrix} \qquad [1.6]$$

Taking into account the *constitutive equation* [1.6], equation [1.4] becomes:

$$\vec{\nabla}^T.\vec{q} = -\begin{bmatrix} \frac{\partial}{\partial x} & \frac{\partial}{\partial y} & \frac{\partial}{\partial z} \end{bmatrix} k \begin{bmatrix} \frac{\partial}{\partial x} \\ \frac{\partial}{\partial y} \\ \frac{\partial}{\partial z} \end{bmatrix} \tau = -\bar{Q} \rightarrow \vec{\nabla}^T.(k\vec{\nabla}\tau) = \bar{Q} \qquad [1.7]$$

This is the second-order equation that the temperature field must satisfy. When the conductivity coefficient is constant, this equation is expressed in terms of the *Laplacian* of the temperature:

$$k\,\Delta\tau = \bar{Q} \quad \text{or} \quad k\left(\frac{\partial^2\tau}{\partial x^2}+\frac{\partial^2\tau}{\partial y^2}+\frac{\partial^2\tau}{\partial z^2}\right)=\bar{Q} \qquad [1.8]$$

$$\begin{aligned} \vec{\nabla}.(k\vec{\nabla}\tau) &= 0 \quad in \quad \Omega \\ \tau &= \bar{\tau} \quad on \quad S_1 \\ \vec{n}\cdot\vec{q} &= \bar{q}_n \quad on \quad S_2 \\ S_1 \cup S_2 &= S \end{aligned}$$

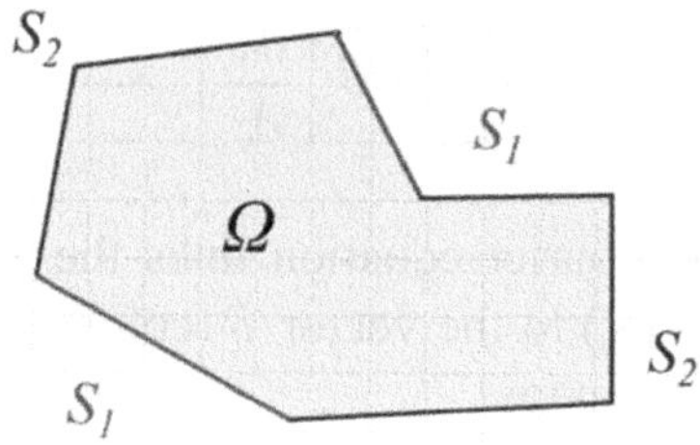

Figure 1.2. *A domain and its boundaries*

The boundary of the domain is divided into two parts:

– S_1 where the temperature is imposed (*Dirichlet* boundary conditions). A bar indicates that the quantity concerned is imposed;

– S_2 where the normal flow is imposed (*Neumann* boundary conditions). If the imposed flows are zero, this part of the boundary is called *adiabatic*.

To calculate the temperature field in a domain Ω, subject to boundary conditions, a system of partial differential equations and the associated boundary conditions must be solved (Figure 1.2). Except in very specific cases, generally in one-dimensional problems, it is not possible to solve this problem analytically.

1.2. Variational method

To solve the system of partial differential equations, it is equivalent to use a variational method and establish the stationarity conditions of the *functional I(τ)* [1.9] expressed in WK (Courant 1943):

$$< I(\tau) = \int_{\Omega} \frac{1}{2} k\ \vec{\nabla}\tau\ .\ \vec{\nabla}\tau\ \text{-}\ \bar{Q}\tau\ d\Omega + \int_{S_2} \bar{q}_n\ \ \tau\ dS > stat. \qquad [1.9]$$

The boundary conditions are taken from Figure 1.2. These are, for example, Dirichlet conditions:

$$\tau = \bar{\tau},\ \delta\tau = 0 \text{ on } S_1 \qquad [1.10]$$

The second term of relation [1.9] serves to introduce Neumann's relations:

$$\vec{n} \cdot \vec{q} = \bar{q}_n \text{ on } S_2 \qquad [1.11]$$

To express the stationarity condition, we write that the difference between the functional calculated with a small arbitrary variation $\delta\tau$ in temperature and the functional [1.9] is equal to zero:

$$I(\tau + \delta\tau) - I(\tau) = 0 \qquad [1.12]$$

Explicitly:

$$\begin{aligned}&\int_{\Omega} \frac{1}{2} k\ \vec{\nabla}(\tau + \delta\tau) \cdot \vec{\nabla}(\tau + \delta\tau)\text{-}\bar{Q}(\tau + \delta\tau)\ d\Omega - \int_{\Omega} \frac{1}{2} k\ \vec{\nabla}\tau \cdot \vec{\nabla}\tau \\ &+\bar{Q}\tau d\Omega + \int_{S_2} \bar{q}_n(\tau + \delta\tau)dS - \int_{S_2} \bar{q}_n \tau dS = 0\end{aligned} \qquad [1.13]$$

We expand this expression:

$$\begin{aligned}&\int_{\Omega} \frac{k}{2} \left\{ \begin{matrix} \vec{\nabla}\tau.\vec{\nabla}\tau + \vec{\nabla}\delta\tau.\vec{\nabla}\delta\tau + \vec{\nabla}\tau.\vec{\nabla}\delta\tau + \vec{\nabla}\delta\tau.\vec{\nabla}\tau \\ -\bar{Q}\tau - \bar{Q}\delta\tau \end{matrix} \right\} d\Omega \\ &-\int_{\Omega} \frac{k}{2}\ \vec{\nabla}\tau \cdot \vec{\nabla}\tau + \bar{Q}\tau d\Omega + \int_{S_2} \bar{q}_n\tau + \bar{q}_n\delta\tau)dS \\ &-\int_{S_2} \bar{q}_n \tau dS = 0\end{aligned} \qquad [1.14]$$

The second term is negligible while terms 1, 5, 7, 8 and 9, 11 cancel each other out. They are therefore removed. Taking into account the commutative property of the scalar product, we obtain:

$$\int_{\Omega} k\ \vec{\nabla}\tau \cdot \vec{\nabla}\delta\tau\ - \bar{Q}\delta\tau d\Omega\ + \int_{S_2} \bar{q}_n\ \delta\tau\ dS = 0 \qquad [1.15]$$

To exploit this result, we extract the variation $\delta\tau$ from the integral using the technique of integration by parts. In this technique, we replace the integration of a product by the sum of the integrals of their derivatives. We start with the simple decomposition:

$$\vec{\nabla}.(f\,\vec{g}) = \vec{\nabla}f.\vec{g} + f\,\vec{\nabla}.\vec{g} \qquad [1.16]$$

In the left-hand side of [1.16], the vector is replaced by the product of the scalar k and the gradient of the function τ, while the scalar f is replaced by the variation $\delta\tau$ of the scalar τ. The relation [1.16] becomes:

$$\vec{\nabla}.\left(\delta\tau\,(k\,\vec{\nabla}\tau)\right) = \vec{\nabla}(\delta\tau)\,.\,(k\vec{\nabla}\tau) + \delta\tau\,\vec{\nabla}.\,(k\vec{\nabla}\tau) \qquad [1.17]$$

The *Gauss–Ostrogradski* theorem expresses the equality between the integral of the divergence of a vector field over a volume and the flux of that field through the boundary of the volume (it is a surface integral). It is applied to the integral on the left-hand side, which gives:

$$\int_{\Omega} \vec{\nabla}.\left(\delta\tau\,k\vec{\nabla}\tau\right)\,d\Omega = \int_{S} \delta\tau\,(k\vec{\nabla}\tau)\,.\,\vec{n}\,dS = \int_{S} \delta\tau\,q_n\,dS \qquad [1.18]$$

In this expression, $\vec{n}$ denotes the outer normal and q_n the normal flux. By integrating [1.18] over the volume and replacing the first term in accordance with [1.18], we obtain:

$$\int_{S} \delta\tau\,q_n dS = \int_{\Omega} \left(\vec{\nabla}(\delta\tau).\,k\vec{\nabla}\tau + \delta\tau\,\vec{\nabla}.\,k\vec{\nabla}\tau\right)\,d\Omega \qquad [1.19]$$

The variation equation is therefore:

$$\int_{\Omega} k\,\vec{\nabla}\tau\cdot\vec{\nabla}\delta\tau\,d\Omega + \int_{S_2} \overline{q}_n\,\delta\tau\,dS = 0 \qquad [1.20]$$

By substituting according to [1.18], we have:

$$\int_{S} \delta\tau q_n dS = -\int_{S_2} \overline{q}_n\,\delta\tau\,dS + \int_{\Omega} \left(\delta\tau\,\vec{\nabla}.\,k\vec{\nabla}\tau\right)\,d\Omega \qquad [1.21]$$

Observing that $\delta\tau = 0$ on S_1, the expression reduces to:

$$\int_{S_2} \delta\tau(q_n - \bar{q}_n)\, dS + \int_{\Omega} \left(\delta\tau\, \vec{\nabla}.\, k\vec{\nabla}\tau\right) d\Omega = 0 \qquad [1.22]$$

By highlighting the variation, we obtain:

$$\begin{pmatrix} \displaystyle\int_{\Omega} \left(\vec{\nabla}.\,(k\,\vec{\nabla}(\tau)) - \bar{Q}\right) d\Omega + \\ \displaystyle\int_{S_2} \left(q_n - \bar{q}_n\right) dS \end{pmatrix} \delta\tau = 0 \qquad [1.23]$$

The variation $\delta\tau$ is arbitrary, and the two terms that multiply it in equation [1.23] have different dimensions, so it is necessary for each of them to cancel out:

$$\begin{aligned} &\vec{\nabla}.\,(k\,\vec{\nabla}(\tau)) = \bar{Q} \text{ in } \Omega \\ &-\vec{n}\cdot(k\,\vec{\nabla}\tau) = \bar{q}_n \text{ on } S_2 \end{aligned} \qquad [1.24]$$

In [1.24], we recognize the constitutive equation and the Neumann boundary condition [1.11]. They are satisfied by the stationarity condition of the functional.

In conclusion, the calculation of the solution to the partial differential equations of the conduction problem (Figure 1.1) can be replaced by the expression of the stationarity conditions of the functional [1.9], which can also be expressed in the following forms:

$$\begin{aligned} &with: D = \begin{bmatrix} \dfrac{\partial\tau}{\partial x} & \dfrac{\partial\tau}{\partial y} & \dfrac{\partial\tau}{\partial z} \end{bmatrix} \cdot \begin{bmatrix} \dfrac{\partial\tau}{\partial x} \\ \dfrac{\partial\tau}{\partial y} \\ \dfrac{\partial\tau}{\partial z} \end{bmatrix} \\ &I(\tau) = \int_{\Omega} \left(\frac{1}{2}kD - \bar{Q}\tau\right) d\Omega + \int_{S_2} \bar{q}_n\, \tau\, dS \end{aligned} \qquad [1.25]$$

1.3. Rayleigh–Ritz method

To find the minimum of [1.25], we use an approximation of the temperature field according to the *Rayleigh–Ritz* method (Ritz 1911; Leissa 2005). The temperature field τ is represented by a combination of n test functions, each assigned a coefficient β_i that will be determined in the minimization process:

$$\tau = \sum_{i=1}^{n} \beta_i f_i(x, y, z) \qquad [1.26]$$

Denoting by $[W]$ the column matrix of the n test functions f_i and by $[B]$ the column matrix of the n parameters β_i, we get:

$$[W] = \begin{bmatrix} f_1 \\ f_2 \\ f_3 \\ \dots \\ f_n \end{bmatrix} \quad [B] = \begin{bmatrix} \beta_1 \\ \beta_2 \\ \beta_3 \\ \dots \\ \beta_n \end{bmatrix} \qquad [1.27]$$

We rewrite [1.26] in a more compact way, in matrix form:

$$\tau = [W]^T [B] \qquad [1.28]$$

The temperature gradient matrix [1.5] (gradient of a scalar) has dimensions 3×1, but if the temperature is expressed as a function of the n parameters contained in $[B]$ [1.28], it becomes a matrix with dimensions $3 \times n$:

$$\vec{\nabla}\tau = \begin{bmatrix} \dfrac{\partial \tau}{\partial x} \\ \dfrac{\partial \tau}{\partial y} \\ \dfrac{\partial \tau}{\partial z} \end{bmatrix} = \begin{bmatrix} \dfrac{\partial}{\partial x}[W]^T \\ \dfrac{\partial}{\partial y}[W]^T \\ \dfrac{\partial}{\partial z}[W]^T \end{bmatrix} [B] = \vec{\nabla}\left([W]^T [B]\right) \qquad [1.29]$$

So, in [1.30] we have the $3 \times n$ matrix of thermal gradients:

$$[G]=\begin{bmatrix} \frac{\partial f_1}{\partial x} & \frac{\partial f_2}{\partial x} & \frac{\partial f_3}{\partial x} & \cdots & \frac{\partial f_n}{\partial x} \\ \frac{\partial f_1}{\partial y} & \frac{\partial f_2}{\partial y} & \frac{\partial f_3}{\partial y} & \cdots & \frac{\partial f_n}{\partial y} \\ \frac{\partial f_1}{\partial z} & \frac{\partial f_2}{\partial z} & \frac{\partial f_3}{\partial z} & \cdots & \frac{\partial f_n}{\partial z} \end{bmatrix} \quad [1.30]$$

The approximate functional $I(\beta)$ derived from [1.25], as a function of these parameters β_i, is written as:

$$\begin{aligned} &< I(\beta) = \frac{1}{2}k[B]^T \int_{\Omega} [G]^T [G] - \bar{Q}[W]^T \, d\Omega \; [B] \\ &+ \int_{S_2} \bar{q}_n \; [\mathrm{W}]^T \, dS[B] > \text{stationary} \end{aligned} \quad [1.31]$$

The size of the central matrix of the gradient product to be integrated is the same as that of the column matrix of parameters $[B]$: n. We have removed the parameters $[B]$ from the first term of the integral, which is a directly calculable matrix:

$$K = \int_{\Omega} k \; [G]^T [G] d\Omega \quad [1.32]$$

If the heat sources and fluxes imposed on boundary S_2 are zero, expression [1.31] simplifies to:

$$\begin{aligned} &< I(\beta) = [B]^T \frac{1}{2} \int_{\Omega} k \; [G]^T [G] d\Omega [B] \\ &= \frac{1}{2}[B]^T \, [K][B] > \text{stationary} \end{aligned} \quad [1.33]$$

The functional of expression [1.9] has been transformed in [1.31] into a quadratic algebraic function of the parameters $[B]$. This expresses that the derivatives of the functional with respect to the coefficients are zero, which, in the case where $q_n = 0$ (zero imposed flows on S_2) and where there are no heat sources, leads to the solution of the linear equation system [1.34]:

$$[K][B]=0 \qquad [1.34]$$

This formalism is extremely attractive and well suited to numerical computation, as it uses only linear equation solving techniques. The only difficulty is that, on the boundary S_1, the approximate solution must satisfy the temperature conditions exactly. Furthermore, since the test functions must cover the entire domain, it is not always easy to find functions that are representative of the exact solution.

1.4. Finite element model

Instead of applying a Rayleigh–Ritz approximation to the entire domain, it is applied to subdomains whose union reconstructs the complete domain. To do this, a certain continuity must be ensured between the elements of this set. In each subdomain or finite element numbered *i*, which is hexahedral in shape, a Rayleigh–Ritz approximation is applied. The temperature field τ_i of element *i* is discretized by trilinear polynomial functions associated with the *nodal temperatures* T_{ij} of the eight vertices (Figure 1.3):

$$\tau_i = \sum_{j=1}^{8} T_{ij} f_j(x,y) \qquad [1.35]$$

To model the entire domain, it is subdivided into hexahedrons, which form a *conforming mesh*: a face shared by two elements has the same nodes, so that the temperature field is continuous.

Figure 1.3 shows the numbering of the nodes of the basic element, whose temperature field is defined in [1.35] and [1.36]. To calculate the *conductivity matrix* of the unit cube element, we start from the definition of the trilinear field:

$$\begin{aligned}\tau_i &= T_1(1-x)(1-y)(1-z) + T_2 x(1-y)(1-z) + T_3\, xy(1-z)\\ &+ T_4(1-x)y(1-z) + T_5(1-x)(1-y)z + T_6\, x(1-y)z\\ &+ T_7\, xyz + T_8(1-x)yz\end{aligned} \qquad [1.36]$$

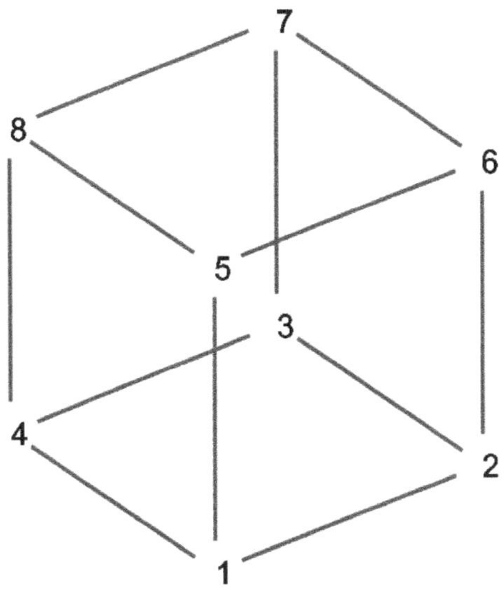

Figure 1.3. *A cube element and its eight vertices*

For $x = 0$, the temperature field of the element is reduced to a bilinear field dependent on the four nodal temperatures 1, 5, 8, 4 of the face:

$$\tau_{x=0} = T_1(1-y)(1-z) + T_4\,y(1-z) + T_5(1-y)z + T_8(1-x)yz \quad [1.37]$$

For the other faces: $x = 1$, $y = 0$, $y = 1$ and $z = 0$, $z = 1$, the results are similar. The six faces of the hexahedron are subject to six bilinear temperature fields. The same applies to any surface of the hexahedron defined by setting one of the three coordinates to a constant value.

Proceeding as in [1.27], we define the *shape functions* $[W_e]$ and the nodal temperatures of the element:

$$\begin{aligned} [W_e] &= \begin{bmatrix} f_1 & f_2 & f_3 & \dots & f_8 \end{bmatrix}^T \\ [T_e] &= \begin{bmatrix} T_1 & T_2 & T_3 & \dots & T_8 \end{bmatrix}^T \end{aligned} \quad [1.38]$$

The shape functions of one of the vertices are shown in Figure 1.4. The temperature gradients are obtained by deriving the temperature field τ_i [1.36]:

$$\vec{\nabla}\tau = \begin{bmatrix} \frac{\partial \tau}{\partial x} & \frac{\partial \tau}{\partial y} & \frac{\partial \tau}{\partial z} \end{bmatrix}^T \quad [1.39]$$

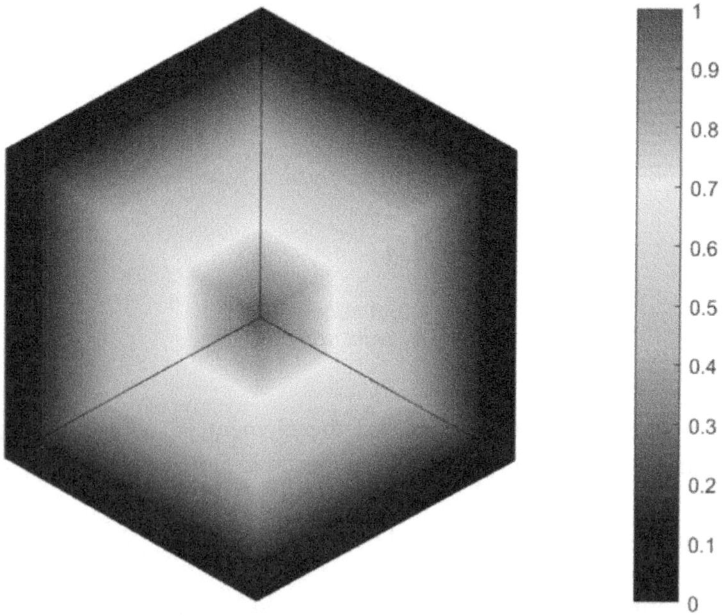

Figure 1.4. *Shape function of one of the vertices of the hexahedron [1.38]*

As in [1.30], we obtain the matrix of gradient operators:

$$[G_e] = \begin{bmatrix} \frac{\partial f_1}{\partial x} & \frac{\partial f_2}{\partial x} & \frac{\partial f_3}{\partial x} & \cdots & \frac{\partial f_8}{\partial x} \\ \frac{\partial f_1}{\partial y} & \frac{\partial f_2}{\partial y} & \frac{\partial f_3}{\partial y} & \cdots & \frac{\partial f_8}{\partial y} \\ \frac{\partial f_1}{\partial z} & \frac{\partial f_2}{\partial z} & \frac{\partial f_3}{\partial z} & \cdots & \frac{\partial f_8}{\partial z} \end{bmatrix} \quad [1.40]$$

Explicitly:

$$[G_e]^T = \begin{bmatrix} -(1-y)(1-z) & -(1-x)(1-z) & -(1-x)(1-y) \\ (1-y)(1-z) & -x(1-z) & -x(1-y) \\ y(1-z) & x(1-z) & -xy \\ -y(1-z) & (1-x)(1-z) & -(1-x)y \\ -(1-y)z & -(1-x)z & (1-x)(1-y) \\ (1-y)z & -xz & x(1-y) \\ yz & xz & xy \\ -yz & (1-x)z & (1-x)y \end{bmatrix} \qquad [1.41]$$

The temperature gradient $\vec{\nabla}\tau$ defined in [1.5] is a 3 × 1 column matrix. The partial derivative matrix $[G_e]$ is 3 × 8, and the column matrix of nodal temperatures $[T_e]$ is 8 × 1:

$$\vec{\nabla}\tau = [G_e][T_e] \qquad [1.42]$$

The conductivity matrix of the element is calculated from the temperature gradient [1.41]:

$$[K_{el}] = \frac{1}{2}\iiint_{\Omega} k[G_e]^T[G_e]dxdydz \qquad [1.43]$$

For a unit cube, it is equal to:

$$[K_{el}] = \frac{k}{12}\begin{bmatrix} 4 & 0 & -1 & 0 & 0 & -1 & -1 & -1 \\ 0 & 4 & 0 & -1 & -1 & 0 & -1 & -1 \\ -1 & 0 & 4 & 0 & -1 & -1 & 0 & -1 \\ 0 & -1 & 0 & 4 & -1 & -1 & -1 & 0 \\ 0 & -1 & -1 & -1 & 4 & 0 & -1 & 0 \\ -1 & 0 & -1 & -1 & 0 & 4 & 0 & -1 \\ -1 & -1 & 0 & -1 & -1 & 0 & 4 & 0 \\ -1 & -1 & -1 & 0 & 0 & -1 & 0 & 4 \end{bmatrix} WK^{-1} \qquad [1.44]$$

The coefficients of the matrix are squares of temperature gradients [1.39] for which nodal temperatures have been identified. They are expressed in m^{-2} and then integrated over the volume of the element. The temperature field is represented by the nodal temperatures [1.38]. The product of the squares of gradients divided by the squares of nodal temperatures and integrated over the volume of the element is therefore expressed in W K^{-1} (W K^{-1} m^{-1} m^{-2} m^3).

The diagonal terms of a cube element are all equal to $k/3$. The non-diagonal terms are equal to 0 if they correspond to nodes directly connected to the diagonal term node (this is the case for three nodes); they are equal to −1 if they are nodes that are not directly connected (this is the case for the remaining four nodes).

The sum of the terms in each row is equal to 0, which implies that, for a constant temperature field, all nodal flows are zero. Similarly, the sum of the terms in each column is equal to 0, which means that the nodal flows balance each other out. The conductivity matrix of the element is therefore singular. Consequently, to solve a problem, at least one nodal temperature must be set. This simple action removes the singularity of the matrix.

For a domain consisting of a single element, the *dissipation function*, which coincides with the functional to be minimized $I(T)$, is constructed from [1.33]. It is equal to:

$$< I(T) = \frac{1}{2}[T_e]^T [K_{el}][T_e] > \text{stationary} \qquad [1.45]$$

The vector (or column matrix $[T_e]$) contains the eight degrees of freedom of the hexahedral finite element. The functional [1.45] does not yet include boundary conditions. To illustrate its use, we seek the temperature field produced by imposing temperatures on the ends of one of the cube's major diagonals. They are 0 K and 30 K, respectively. The problem is solved explicitly. After being reordered, the conductivity matrix of the element [1.44] is subdivided into four submatrices of dimensions 6 × 6, 6 × 2, 2 × 6, and 2 × 2:

$$K_{el} = \begin{bmatrix} K_{ii} & K_{is} \\ K_{si} & K_{ss} \end{bmatrix} \qquad [1.46]$$

We solve the system of equations [1.46]. In the conductivity matrix of [1.46], the conductivity k = 12 W m^{-1} K^{-1}.

$$\begin{bmatrix} K_{ii} & K_{is} \\ K_{si} & K_{ss} \end{bmatrix} \begin{bmatrix} T \\ \bar{T} \end{bmatrix} = \begin{bmatrix} 0 \\ R \end{bmatrix} \begin{array}{l} \rightarrow \; [T] = -[K_{si}]^{-1}[K_{is}][\bar{T}] \\ \rightarrow \; [R] = [K_{si}][T]+[K_{si}][\bar{T}] \end{array} \qquad [1.47]$$

The solution to [1.47] is $[T]^T = [T^T \; \bar{T}^T] = [10\ 20\ 10\ 20\ 20\ 10\ 0\ 30]\,K$. These are the temperatures of nodes [7 2 3 4 5 6 8 1] (Figure 1.3). The second member of relation [1.47] is $[R]^T = [-60\ 60]\ W$.

The surface temperature maps of the cube are shown in Figures 1.5 and 1.6.

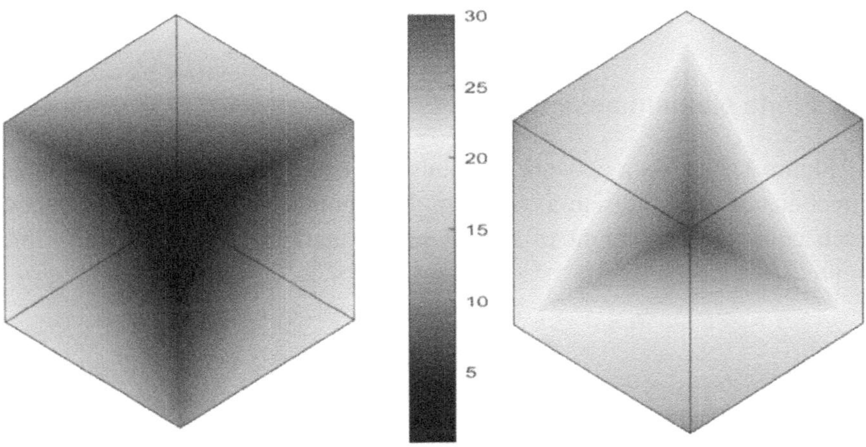

Figure 1.5. *Isotherms on the surface of the cube subjected to temperatures of 0 and 30 K*

Dissipation is calculated according to formula [1.45]:

$$\frac{1}{2}[T]^T [K][T] = 900 \text{ WK} \qquad [1.48]$$

These results are also presented in such a way that the two peaks of maximum and minimum temperature are simultaneously visible on one or both sides (Figure 1.6).

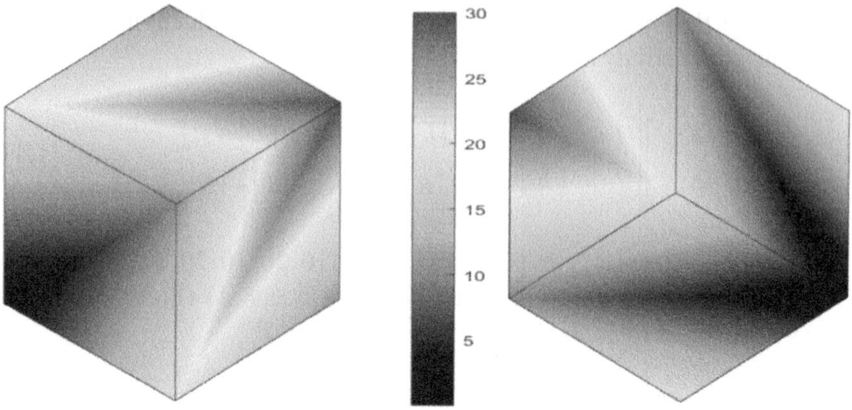

Figure 1.6. *Isotherms with peaks at 0 and 30 K simultaneously visible*

1.5. Finite element assembly

Figure 1.7 shows a network of points spaced at unit intervals in the three directions of the Cartesian axes. From these points, we define $N = 8$ cubes. Each cube connects eight of the 27 points marked in black in the same figure.

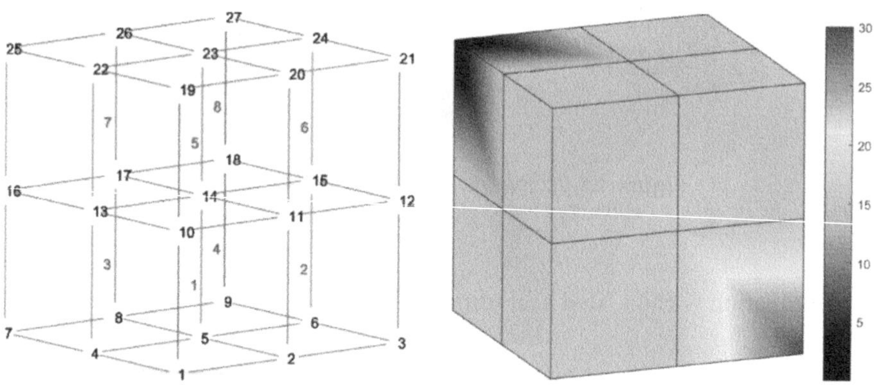

Figure 1.7. *Eight cubes, temperatures imposed at the ends of a large diagonal*

To verify the continuity of the temperature field across the domain, the temperature field is identical on each face separating two elements, which means that the nodal temperatures of elements sharing the same node are the same at that node. In the mesh in Figure 1.7, for example, the third node of element 3 and the fourth node of element 4 are assigned to global node 8. The localization matrix [*L*] in Table 1.1 is 8 × 8 in size and defines the localizations of the eight cubes in the analyzed domain.

Element number: *n*	Numbering of the eight nodes of the elements in the global system: localization matrix [*L*]
1	1 2 5 4 10 11 14 13
2	2 3 6 5 11 12 15 14
3	4 5 8 7 13 14 17 16
4	5 6 9 8 14 15 18 17
5	10 11 14 13 19 20 23 22
6	11 12 15 14 20 21 24 23
7	13 14 17 16 22 23 26 25
8	14 15 18 17 23 24 27 26

Table 1.1. *Localization matrix [L] of the eight elements in Figure 1.7*

In the domain composed of eight cube elements (Figure 1.7), the contributions of the elements are added together to form the global discretized functional *I* (*T*) [1.50]. The column matrix [*T*] denotes the nodal temperatures numbered, in this case, from 1 to 27. The index *n* corresponds to the element number, while the indices *i* and *j* designate the position of the node in the sequence of eight nodes of an element. These are column indices of the matrix [*L*] in Table 1.1. The eight shape functions of each element are defined in [1.36], and the 24 gradients in [1.40]. The equivalent for the finite element mesh of the functional [1.33] is expressed as a function of the elementary nodal temperatures *T* (*L* (*n*, *i*)).

The single-column matrix *T* contains the 27 nodal temperatures of the finite element mesh:

$$\sum_{n=1}^{N}\sum_{i=1}^{8}\sum_{j=1}^{8}\left(\int_{\Omega_n}\frac{1}{2}k_n\,\left[T(L(n,i))\right]^T\left[G_n\right]^T\left[G_n\right]\left[T(L(n,j))\right]\,d\Omega_n\right) = \frac{1}{2}[T]^T\left[K_g\right][T] \qquad [1.49]$$

The matrix $[K_g]$ is the global conductivity matrix. Figure 1.7 shows that the second node of element 1 is the same as the first node of element 2. Similarly, node 11 is node 6 of element 1, node 5 of element 2, node 2 of element 5 and node 1 of element 6.

The correspondences between elementary nodes and global nodes are listed for the eight elements of the domain in the localization matrix $[L]$ in Table 1.1. Thus, in this same table, we can see that node 14, located at the center of the domain, is present in all eight elements.

The global conductivity matrix is the sum of the elementary matrices positioned at the nodes defined in Table 1.1. In [1.48], the dissipation function was obtained:

$$< I(T) = \frac{1}{2}[T]^T \left[K_g\right][T] > \qquad [1.50]$$

EXAMPLE.– Let us revisit the cube problem by refining the mesh: from a single element, we move to a 2 × 2 × 2 mesh (Figure 1.7). The temperatures of nodes 1 and 27 are set to 30 and 0 K. We find a dissipation of 1/2 $([T]^T [K]\ [T]) = 429\ WK$ and flows at the two imposed nodes equal to $\pm 28.6\ W$. We also verify that half of the sum of the actions of the flows on the imposed temperatures is equal to the dissipation: $28.6\ W \times 30\ K/2 = 429\ WK$.

1.6. Convergence of the finite element method

The decomposition of a finite element into sub-elements that reproduce the initial domain and its boundary conditions leads to the study of the convergence of the finite element method. The aim is to see how refining a mesh improves the solution.

Figure 1.8 shows the convergence results for meshes with 4, 121 and 961 nodes. In the case of four nodes, the temperature field is visualized continuously or using isolines.

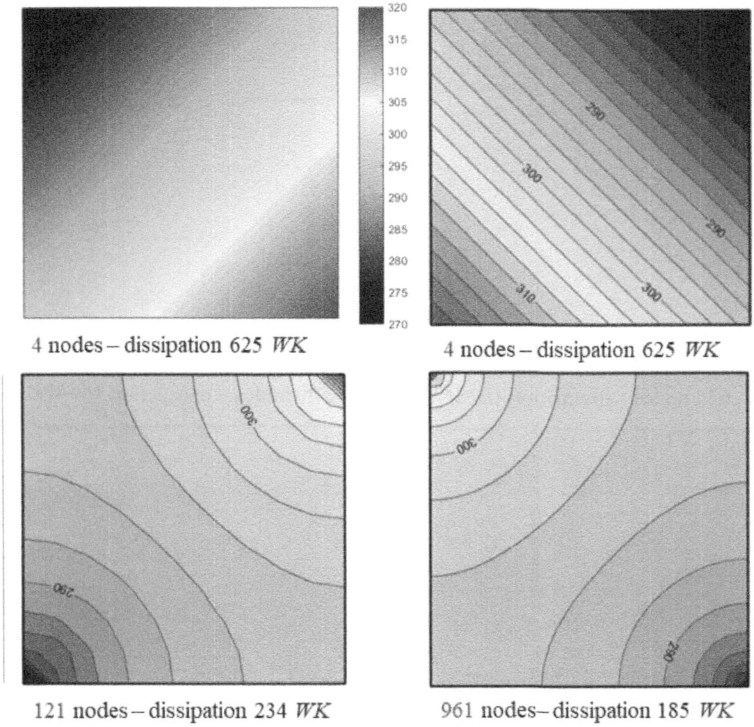

Figure 1.8. *Square with temperatures imposed at the ends of a diagonal*

The quality of the finite element solution to a steady-state heat conduction problem depends on the quality of the mesh (Figure 1.9). It is easy to imagine that a greater number of elements are needed where the temperature varies more significantly. To achieve this, the mesh must be refined in the areas affected by these significant variations. There are methods for determining the quality of the solution and the degree of refinement required.

The first indicator of the quality of the solution is the value of the dissipation function [1.49]. In problems involving only imposed temperatures and adiabatic boundaries, it has been shown (Fraeijs de Veubeke and Hogge 1972) that dissipation is an upper bound on the exact solution, in other words:

$$I(T)=\frac{1}{2}[T]^T\left[K_g\right][T]>I_{exact} \qquad [1.51]$$

625 nodes – dissipation 192 *WK*

2,500 nodes – dissipation 169 *WK*

8,625 nodes – dissipation 158 *WK*

10,000 nodes – dissipation 151 *WK*

Figure 1.9. *Convergence, temperatures imposed at the ends of a diagonal*

The results in Figure 1.10 show that dissipation decreases from 625 *WK* for 4 nodes to 185 *WK* for 961 nodes, and in Figure 1.9, dissipation decreases to 151 *WK* for 10,000 nodes.

In the problem with 25 elements and 36 nodes (Figure 1.11), the dissipation is 281 *WK*. When the temperature gradients are visualized using arrows whose length is proportional to their modulus, the result is consistent with the isotherm plot (Beckers and Beckers 2016a). The maximum gradient is 12 Km^{-1} and the average is 2.4 Km^{-1}. Near the edges, the arrows are oriented like the edges and are longer where the isotherms are closer together, namely, at the two vertices with imposed temperatures.

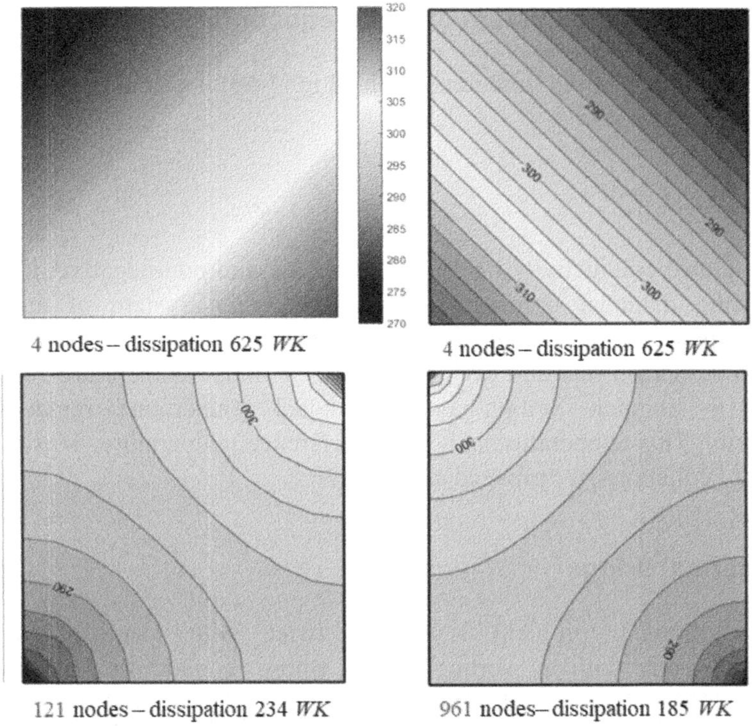

Figure 1.10. *Convergence study*

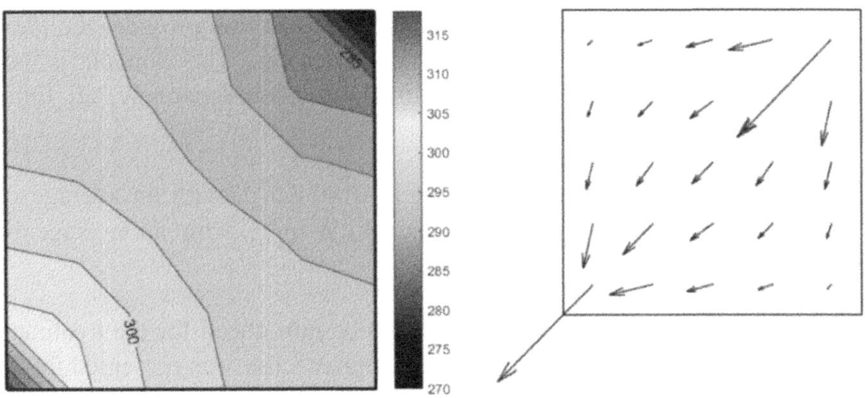

Figure 1.11. *Gradient visualization, 36 nodes, 281 WK dissipation*

1.7. Solution to the system of linear equations

The minimum of the functional defined in [1.50] is calculated by solving the system of equations:

$$\left[K_g\right][T]=0 \qquad [1.52]$$

The global conductivity matrix is symmetric and non-negative; it is also sparse and diagonal dominant, which makes the system of equations particularly well-conditioned. The sum of the coefficients of the matrix is equal to 0 on each row and each column. If no nodal temperature is set, the system is singular. Setting a single nodal temperature removes the singularity. This temperature acts as the reference temperature. To solve the system of equations, we proceed as in [1.47].

1.8. Thermal bridge

The following problem is treated twice with identical Dirichlet conditions, but with a vertical band whose conductivity is normal: 1 W m^{-1} K^{-1} (Figure 1.12(a)) or very high: 10,000 W m^{-1} K^{-1} (Figure 1.12(b)).

The thermal bridge is represented in Figure 1.13 by a vertical yellow band where the conductivity is 10,000 times greater than in the rest of the domain. Visualizing temperature gradients and heat flows provides the ideal graphical means of representing this phenomenon. The domain is not homogeneous in terms of materials. However, taken separately, all finite elements are homogeneous.

The maximum temperature gradient is –140 Km^{-1}, with an average of –13 Km^{-1}. The maximum heat flux is 220 W m^{-2}, with an average of 35 W m^{-2}.

An *anisotropic* constitutive law can also be introduced for the treatment of this type of material. In equation [1.6], Fourier's law was presented for an isotropic body, in which the relationship between heat flux and temperature gradient remains the same regardless of the orientation of these quantities. However, more complex situations can be considered, in which the heat flux is no longer proportional to the temperature gradient, but is related to it by a tensor. We can then rewrite [1.6] in indicial notation:

$$q_j = k_{ji} \partial_i \tau \qquad [1.53]$$

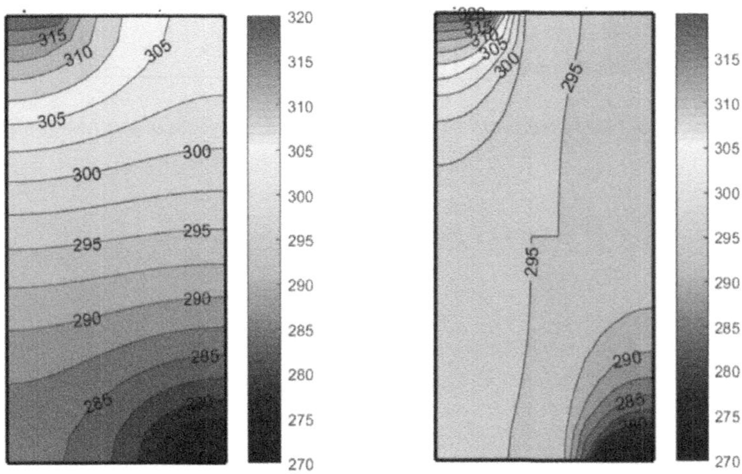

Dissipation 390 WK Dissipation 741 WK

a) b)

Figure 1.12. *Isotherms with a vertical thermal bridge on the right*

The term k_{ij} represents the symmetric thermal conductivity tensor (Fraeijs de Veubeke et al. 1972). When the principal axes of this tensor have a constant orientation from point to point and if the Cartesian axes are parallel to them, we have a simpler law:

$$q_1 = -k_1 \partial_1 \tau \, ; \, q_2 = -k_2 \partial_2 \tau \, ; \, q_3 = -k_3 \partial_3 \tau \qquad [1.54]$$

This relationship corresponds to materials known as *orthotropic*.

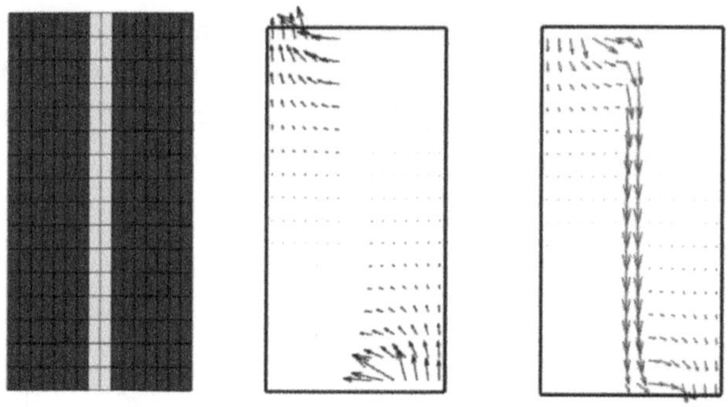

Figure 1.13. *Heat flows and temperature gradients, vertical thermal bridge*

2

Boundary Conditions

One of the many advantages of the finite element method is the ease with which boundary conditions can be introduced. For Dirichlet conditions, it is sufficient to fix the relevant nodal variables. To facilitate use, the *linear constraint* technique is introduced, which allows non-conforming meshes to be glued. For Neumann boundary conditions, the variables conjugate to the nodal temperatures are defined. These are *generalized fluxes*, which constitute the second members of the equations to be solved. When thermal fluxes are related to what is happening outside the mesh, it is necessary to analyze radiation interactions or interactions with a fluid. This case is analyzed by introducing special elements called convection elements. Radiation will be discussed in the second part of the book.

2.1. Dirichlet boundary conditions

At least one Dirichlet boundary condition must be imposed, and then the values of the non-imposed (or free) nodal temperatures are calculated. To solve the system of equations, the nodal temperature vector is divided into two groups: the free unknown temperatures $[T_1]$ and the imposed temperatures $[T_2]$. The system is written as:

$$\begin{bmatrix} K_{11} & K_{12} \\ K_{21} & K_{22} \end{bmatrix} \begin{bmatrix} T_1 \\ T_2 \end{bmatrix} = 0 \qquad [2.1]$$

For a color version of all the figures in this chapter, see www.iste.co.uk/beckers/finite.zip.

Its solution is:

$$[T_1] = [K_{11}]^{-1}[K_{12}][T_2] \quad [2.2]$$

To obtain this distribution of free and fixed nodes, in the example of Figure 1.13, the nodes on the diagonal supporting the fixations at one end of the overall conductivity matrix must be grouped together, as in [1.46].

To fix other nodes, it is necessary to change the numbering in order to group this new list at the end of the global matrix. This is not very practical.

2.1.1. *Linear constraint technique*

This method is based on the *Lagrange multiplier* technique (Quiroz and Beckers 1995). It is a method for finding the stationary points (maximum, minimum, etc.) of a differentiable function of one or more variables under constraints. Suppose we need to find an *extremum* of a function $f(x)$ subject to the condition:

$$g(x) = c \quad [2.3]$$

To solve this problem, we are looking for an *extremum* of the augmented function:

$$f(x) + \lambda(g(x) - c) \quad [2.4]$$

This research leads to the solution of Euler's equations, which consist of canceling the derivatives with respect to the variables x and λ. With respect to x, we have:

$$\frac{df}{dx} + \lambda\frac{dg}{dx} = 0 \quad [2.5]$$

Canceling the derivative of [2.4] with respect to the multiplier λ restores constraint [2.3].

We therefore choose to formulate Dirichlet's conditions using linear relations and Lagrange multipliers. This amounts to explicitly writing the values of the fixed temperatures:

$$[N]\,[T]=\left[\bar{T}\right] \qquad [2.6]$$

For each fixation, we have a constraint that is linear, because the coefficients of the matrix N are constants. The set of linear constraints is added to the functional with Lagrange multipliers [2.4].

In expression [2.6], $[T]$ is the column matrix of nodal temperatures and $[N]$ is a matrix with as many rows as fixed nodes and the same number of columns as nodal variables (size of $[T]$). Each row of the linear constraint matrix $[N]$ contains only zeros and a 1 in the column corresponding to the fixed node. As in [2.4], we write the augmented function with the column matrix $[\Lambda]$ of Lagrange multipliers:

$$< I(T)=[T]^T[K][T]+[\Lambda]^T\left([N][T]-\left[\bar{T}\right]\right)> extremum \qquad [2.7]$$

Euler's equations for minimizing this functional are obtained by differentiating it with respect to $[\Lambda]$ and $[T]$:

1) the derivative with respect to the Lagrange multipliers $[\Lambda]$ restores equation [2.6];

2) the derivative with respect to $[T]$ yields:

$$[K][T]+[N]^T[\Lambda]=[0] \qquad [2.8]$$

Combining [2.6] and [2.8] yields the system:

$$\begin{bmatrix} K & N^T \\ N & 0 \end{bmatrix}\begin{bmatrix} T \\ \Lambda \end{bmatrix}=\begin{bmatrix} 0 \\ \bar{T} \end{bmatrix}\rightarrow[A][B]=[G] \qquad [2.9]$$

The column matrix $[G]$ of the second member contains a series of terms whose number corresponds to the size of $[T]$. These terms are followed by the imposed temperatures $[\bar{T}]$. The unknown vector $[B]$ contains the set of nodal temperatures and Lagrange multipliers that represent the generalized heat fluxes passing through the fixed nodes. Unlike the matrix $[K]$, the matrix $[A]$ is non-singular, provided that the constraints are independent of each other.

In the example in Figure 2.1, with a conduction coefficient $k = 12$ W m^{-1} K^{-1} and a temperature difference of 100 K, the dissipation [1.52] is equal to 17,877.8348 *WK*. At the ends of the domain, the heat fluxes are 357.5567 *W*.

Explicitly, the Dirichlet boundary conditions are T (9) = 200 K, T (10) = 200 K, T (19) = 200 K, T (20) = 200 K, T (4) = 100 K, T (8) = 100 K, T (14) = 100 K and T (18) = 100 K. To solve system [2.7], eight rows and eight columns must therefore be added to the matrix.

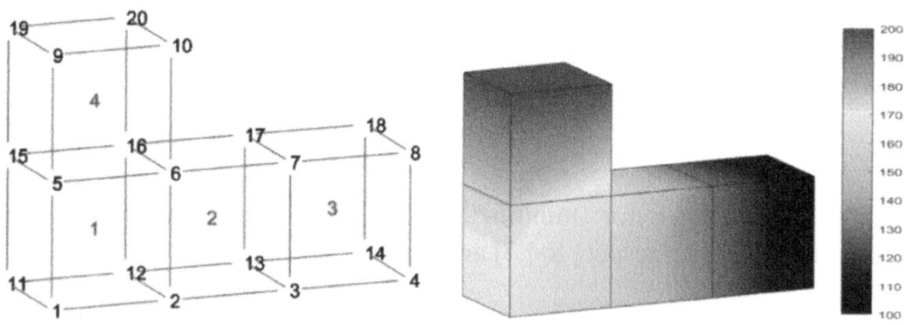

Figure 2.1. *Temperatures imposed on two faces at the ends of the polyhedron*

The linear constraint method can also be used to solve the problem by imposing an average temperature of 200 K on the first face and 100 K on the second. Only two rows and two columns need to be added to the matrix. The results are slightly different from the previous case. The dissipation is now equal to 17,878 *WK*. At the ends of the domain, the heat flows are identical in absolute value at all nodes affected by the constraints. Their absolute values are all equal to 89 *W*. As the system is less constrained than in the previous case, the dissipation and reactive flows are slightly lower in absolute value. For the isotherms, there is no difference from Figure 2.1.

2.1.2. *Mesh gluing*

The mesh in Figure 2.2 is not compatible. This is because the upper element is a half-cube. Two of its lower nodes (nodes 21 and 22) are located in the middle of the edges of the cube below.

The gluing operation is performed by specifying that the temperature of node 21 is equal to half the sum of the temperatures of nodes 5 and 6, and that the temperature of node 22 is equal to half the sum of the temperatures of nodes 15 and 16. There are 10 linear relationships to be respected: eight for the fixations and two for the gluing. The temperature is set at 100 K on the vertical square face and 200 K on the upper horizontal face.

With a conduction coefficient $k = 12$ W m^{-1} K^{-1}, the dissipation is equal to 12,692.2461 *WK*. At the ends of the domain, on the half-square of the upper horizontal face and the vertical square on the right, the heat fluxes are equal to 254 *W*. The product of one of these fluxes by half of the imposed temperature difference of 100 K is equal to the dissipation (253.8449 $W \times 50\ K = 12{,}692\ WK$).

When the two cubes at the bottom (cubes 2 and 3) are replaced by a rectangular parallelepiped, the dissipation changes very little, going from 12,692 *WK* to 12710 *WK* and the heat flux from 253.8 *W* to 254.2 *W*.

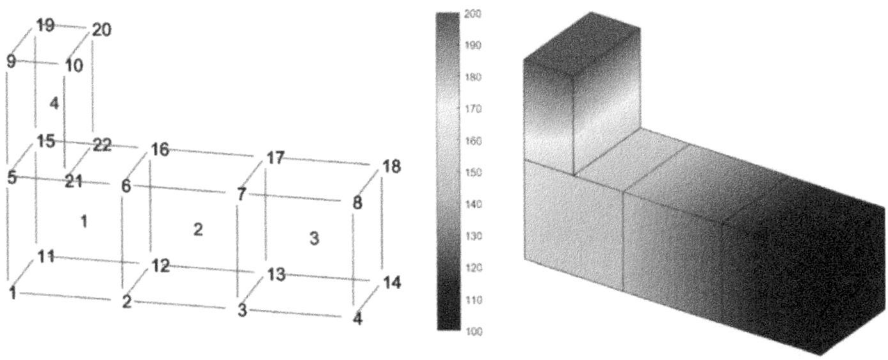

Figure 2.2. *Temperatures imposed on two faces and mesh gluing*

2.2. Neumann boundary conditions

In Chapter 1, the conducting medium was always bounded by an adiabatic surface, except where temperatures were imposed. In many situations, however, we want to be able to impose heat fluxes on the S_2 part of the domain boundary:

$$< I(\tau) = \int_{\Omega} \frac{1}{2} k \, \vec{\nabla}\tau \, . \, \vec{\nabla}\tau \, d\Omega + \int_{S_2} \overline{q}_n \; \tau \, dS > \text{ stationary} \quad [2.10]$$

From this, we can deduce von Neumann's boundary conditions [2.8]:

$$-\vec{n} \cdot (k \, \vec{\nabla}\tau) = \overline{q}_n \text{ on } S_2 \quad [2.11]$$

The temperature of a face belonging to element i is expressed as a function of the four temperatures of the vertices of the face and bilinear shape functions [1.37]:

$$\tau_f = T_{i1}\left(1-\frac{x}{a}\right)\left(1-\frac{y}{b}\right) + T_{i2}\frac{x}{a}\left(1-\frac{y}{b}\right) + T_{i3}\frac{x}{a}\frac{y}{b} + T_{i4}\left(1-\frac{x}{a}\right)\frac{y}{b} \quad [2.12]$$

Limiting the demonstration to one face of the element, we can write that the second term of equation [2.11] corresponds to the sum of the products of the generalized nodal flows g_i (W) and the nodal temperatures T_i:

$$\int_{S_{edge}} \overline{q}_n \tau_f \, dS_f = g_1 T_{fa1} + g_2 T_{fa2} + g_3 T_{fa3} + g_4 T_{fa4} \quad [2.13]$$

We discretize the functional:

$$\int_{S_f} \overline{q}_n \tau dS_f = \int_{S_{edge}} \overline{q}_n \begin{pmatrix} T_{i1}\left(1-\frac{x}{a}\right)\left(1-\frac{y}{b}\right) + T_{i2}\frac{x}{a}\left(1-\frac{y}{b}\right) \\ +T_{i3}\frac{x}{a}\frac{y}{b} + T_{i4}\left(1-\frac{x}{a}\right)\frac{y}{b} \end{pmatrix} dS_f \quad [2.14]$$

In matrix form:

with: $[T]=[T_1 \quad T_2 \quad T_3 \quad T_4]^T$; we have:

$$\int_{S_{2edge}} \bar{q}_n \tau dS_f = \int_{S_{2edge}} \bar{q}_n \begin{bmatrix} \left(1-\frac{x}{a}\right)\left(1-\frac{y}{b}\right)T_1 + \frac{x}{a}\left(1-\frac{y}{b}\right)T_2 \\ +\frac{x}{a}\frac{y}{b}T_3 + \left(1-\frac{x}{a}\right)\frac{y}{b}T_4 \end{bmatrix} dS_f \qquad [2.15]$$

We can extract the temperatures from the integral:

$$\int_{S_{2edge}} \bar{q}_n \tau dS_{face} =$$

$$\int_{S_{2edge}} \bar{q}_n \left[\left(1-\frac{x}{a}\right)\left(1-\frac{y}{b}\right) \quad \frac{x}{a}\left(1-\frac{y}{b}\right) \quad \frac{x}{a}\frac{y}{b} \quad \left(1-\frac{x}{a}\right)\frac{y}{b}\right] dS_{face} \begin{bmatrix} T_1 \\ T_2 \\ T_3 \\ T_4 \end{bmatrix} \qquad [2.16]$$

Finally, the second imposed member is written in the form:

$$[F_1 \quad F_2 \quad F_3 \quad F_4] =$$

$$\int_{S_f} \bar{q}_n \left[\left(1-\frac{x}{a}\right)\left(1-\frac{y}{b}\right) \quad \frac{x}{a}\left(1-\frac{y}{b}\right) \quad \frac{x}{a}\frac{y}{b} \quad \left(1-\frac{x}{a}\right)\frac{y}{b}\right] dS_f \qquad [2.17]$$

To illustrate the use of Neumann boundary conditions, we use the same example as in Figure 2.1, but on the upper horizontal face we replace the Dirichlet conditions (imposed temperature of 200 K) with Neumann conditions (imposed flux: $\bar{q}_n = 360$ W m^{-2}).

With a conductivity coefficient of 12 W m^{-1} K^{-1}, the dissipation is equal to 18,123 *WK* (Figure 2.3). The average temperature of the upper surface is 200.7 K.

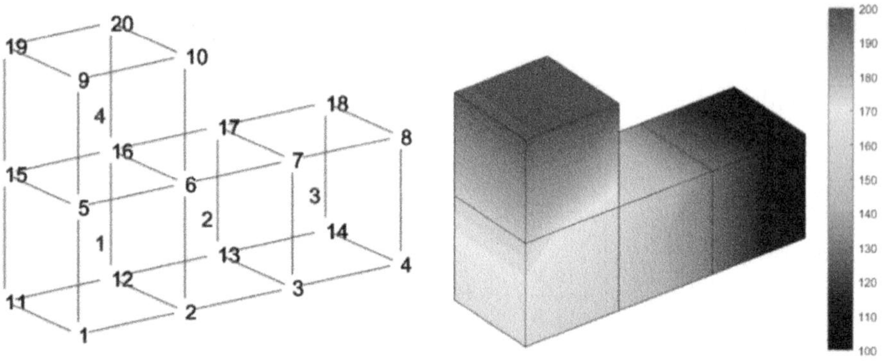

Figure 2.3. *Dirichlet on the right vertical face, Neumann on the top face*

2.3. Convection

2.3.1. *Formulation of the convection element*

When the conductive domain is immersed in a fluid, whether liquid or gaseous, heat exchange is considered to occur with the fluid as a whole.

The partial differential equations for the problem of heat transfer by convection are derived from the steady-state conditions of the functional [2.18].

The quantity S_3 represents all faces of the solid subject to convection conditions. Recall that S_2 corresponds to the part of the boundary subject to Neumann conditions:

$$< I(\tau) = \int_{\Omega} \frac{1}{2} k \, \vec{\nabla}\tau \, . \, \vec{\nabla}\tau \, d\Omega + \frac{1}{2} \int_{S_3} h\left(\tau - \tau_f\right)^2 dS + \int_{S_2} \overline{q}_n \ \tau dS > \min \quad [2.18]$$

The Rayleigh–Ritz procedure is the same as in the pure conduction problem. We can therefore directly examine how to calculate the "conductivity" matrices of the convective elements by writing the functional at the level of a convective face:

$$I_{el} = \frac{1}{2} \int_{S_f} h\left(\tau - \tau_f\right)^2 dS_f \quad [2.19]$$

In [2.18] and [2.19], the quantity h represents the *convection coefficient* measured in W K^{-1} m^{-2}. The temperature τ_f of the fluid is assumed to be uniform. We are studying an element of the boundary of a solid that we call the skin. If it is the skin of a single face of a volume element, it is a quadrilateral, therefore with straight edges, defined by its four vertices $[Q]$:

$$[Q]=\begin{bmatrix} Q_1 \\ Q_2 \\ Q_3 \\ Q_4 \end{bmatrix}=\begin{bmatrix} x_1 & y_1 & z_1 \\ x_2 & y_2 & z_2 \\ x_3 & y_3 & z_3 \\ x_4 & y_4 & z4 \end{bmatrix}=\begin{bmatrix} X & Y & Z \end{bmatrix} \quad [2.20]$$

The matrix $[Q]$, with four rows and three columns, contains the Cartesian coordinates of the four vertices (points or nodes that define the element). The points $[P]$ belonging to the element are defined by shape functions, therefore:

$$P(s,t)^T = \left[x(s,t)\,y(s,t)\,z(s,t)\right]=[F]Q \quad [2.21]$$

More explicitly, the transpose of the point vector is a row matrix that is a function of two parameters. The shape functions f_i of the four vertices are functions equal to 1 at one of the vertices and 0 at the others. They are shown in Figure 2.4 with different discretizations: 1 × 1 for the lower right vertex, 2 × 2 for the lower left vertex, 4 × 4 for the upper right vertex, and 8 × 8 for the upper left vertex. Apart from a rotation of approximately 90°, these four functions are identical, but their graphical representations differ because, in Figure 2.4, more or less fine discretizations have been used:

$$\begin{aligned}&[F]=[f_1 \quad f_2 \quad f_3 \quad f_4]=\\&\left[(1-s)(1-t) \quad s(1-t) \quad st \quad (1-s)t\right]\end{aligned} \quad [2.22]$$

The set of functions f_i forms a barycentric combination satisfying the *partition of unity* condition:

$$f_1+f_2+f_3+f_4=1 \quad [2.23]$$

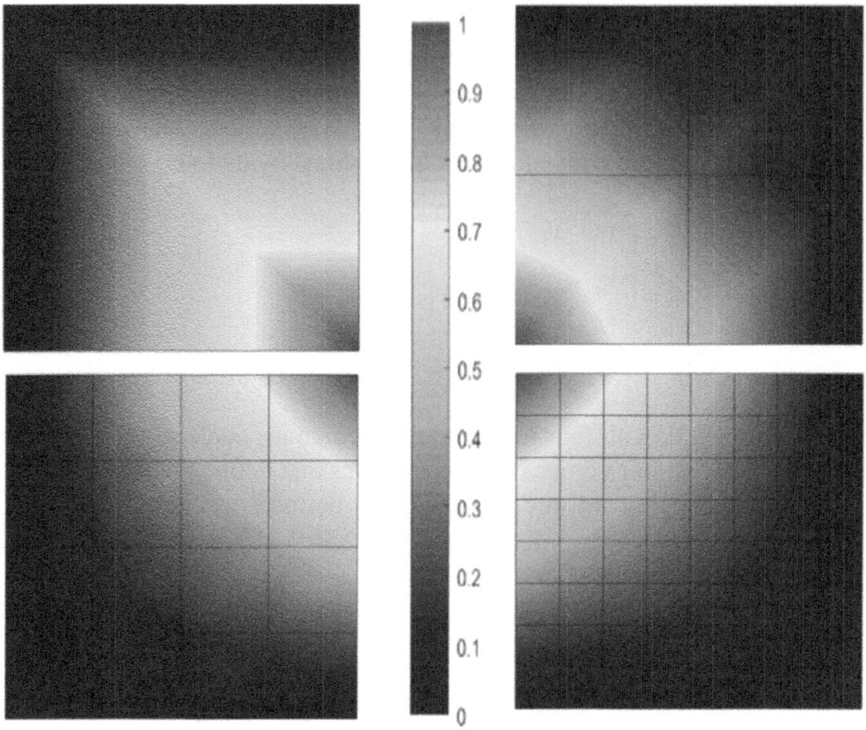

Figure 2.4. *Square shape function*

The position of any point on the element is expressed in terms of the coordinates of the four vertices [*Q*] and the shape functions *f (s, t)* stored in vector [*F*] [2.24]:

$$
\begin{aligned}
x(s;t) &= [F][X] = \begin{bmatrix}(1-s)(1-t) & s(1-t) & st & (1-s)t\end{bmatrix}[X] \\
y(s;t) &= [F][Y] = \begin{bmatrix}(1-s)(1-t) & s(1-t) & st & (1-s)t\end{bmatrix}[Y] \\
z(s;t) &= [F][Z] = \begin{bmatrix}(1-s)(1-t) & s(1-t) & st & (1-s)t\end{bmatrix}[Z]
\end{aligned}
\quad [2.24]
$$

The derivatives of the Cartesian coordinates x and y [2.24] expressed in parametric coordinates s and t are:

$$\begin{aligned}
\frac{\partial x(s,t)}{\partial s} &= \frac{\partial[F]}{\partial s}[X] = \begin{bmatrix}-(1-t) & (1-t) & t & -t\end{bmatrix}[X] \\
\frac{\partial x(s,t)}{\partial t} &= \frac{\partial[F]}{\partial t}[X] = \begin{bmatrix}-(1-s) & -s & s & (1-s)\end{bmatrix}[X] \\
\frac{\partial y(s,t)}{\partial s} &= \frac{\partial[F]}{\partial s}[Y] = \begin{bmatrix}-(1-t) & (1-t) & t & -t\end{bmatrix}[Y] \\
\frac{\partial y(s,t)}{\partial t} &= \frac{\partial[F]}{\partial t}[Y] = \begin{bmatrix}-(1-s) & -s & s & (1-s)\end{bmatrix}[Y] \\
\frac{\partial z(s,t)}{\partial s} &= \frac{\partial[F]}{\partial s}[Z] = \begin{bmatrix}-(1-t) & (1-t) & t & -t\end{bmatrix}[X] \\
\frac{\partial z(s,t)}{\partial t} &= \frac{\partial[F]}{\partial t}[Z] = \begin{bmatrix}-(1-s) & -s & s & (1-s)\end{bmatrix}[X]
\end{aligned} \qquad [2.25]$$

The Jacobian matrix is written as:

$$\begin{aligned}
[J] &= \left[\frac{\partial P}{\partial s} \times \frac{\partial P}{\partial t}\right] = \\
&\left[\frac{\partial y}{\partial s}\frac{\partial z}{\partial t} - \frac{\partial y}{\partial t}\frac{\partial z}{\partial s} \quad \frac{\partial z}{\partial s}\frac{\partial x}{\partial t} - \frac{\partial x}{\partial s}\frac{\partial z}{\partial t} \quad \frac{\partial x}{\partial s}\frac{\partial y}{\partial t} - \frac{\partial x}{\partial t}\frac{\partial y}{\partial s}\right]
\end{aligned} \qquad [2.26]$$

Its determinant is called the *Jacobian*:

$$\begin{aligned}
&\det\left(\left[J(s,t)\right]\right) = \\
&\sqrt{\left(\frac{\partial y}{\partial s}\frac{\partial z}{\partial t} - \frac{\partial y}{\partial t}\frac{\partial z}{\partial s}\right)^2 + \left(\frac{\partial z}{\partial s}\frac{\partial x}{\partial t} - \frac{\partial x}{\partial s}\frac{\partial z}{\partial t}\right)^2 + \left(\frac{\partial x}{\partial s}\frac{\partial y}{\partial t} - \frac{\partial x}{\partial t}\frac{\partial y}{\partial s}\right)^2}
\end{aligned} \qquad [2.27]$$

It is rarely possible to express a function $g(s, t)$ in Cartesian coordinates: $g(s, t) \rightarrow g^*(x, y)$. The integral over the element of the function g depending on the two parameters s and t is identical to the integral of its equivalent in Cartesian coordinates:

$$\iint_{\Omega_{Coons}} g^{*}(x,y)\,dx\,dy = \int_0^1\int_0^1 g(s,t)J(s,t)\,ds\,dt \qquad [2.28]$$

The above integral is evaluated numerically, because if the shape of the element is arbitrary, its Jacobian cannot be easily calculated analytically.

In parametric coordinates s and t, both of which vary between 0 and 1, the temperature of the face of a solid element is discretized as follows:

$$\tau = \left[(1-s)(1-t) \quad s(1-t) \quad st \quad (1-s)t\right][T] = \lfloor F_c \rfloor [T] \qquad [2.29]$$

$$\tau_i = \sum_{j=1}^{4} T_{ij} f_{ij}(x,y) \qquad [2.30]$$

This expression is introduced into [2.19]:

$$I_{el} = \frac{1}{2}\int_{S_f} h\left(\lfloor F_c \rfloor [T] - \tau_f\right)^2 dS_f \qquad [2.31]$$

By explaining and expanding, we obtain:

$$I_{el} = \frac{1}{2} h \int_{S_f} \left\{[T]^T \lfloor F_c \rfloor^T \left(\lfloor F_c \rfloor [T] - 2\lfloor F_c \rfloor [T]\tau_f + \tau_f^2\right)\right\} S_f \qquad [2.32]$$

A new notation is used which includes the fluid temperature in the list of nodal temperatures, with [2.33] using the notation $\tau_f = T_f$ in which the fluid temperature is equated with that of a *virtual node*:

$$[T_{el}] = \left[T_1 \quad T_2 \quad T_3 \quad T_4 \quad T_f\right]^T \qquad [2.33]$$

$[T_{el}]$ is the vector containing the nodal temperatures of a convection element:

$$I_{el} = \frac{1}{2} h T_{el}^T \left(\int_{S_f} \begin{bmatrix} [F_c]^T \\ -1 \end{bmatrix} \left[[F_c] - 1 \right] df \right) T_{el} \qquad [2.34]$$

Taking into account that the temperature is defined in parametric coordinates:

$$I_{el} = \frac{1}{2} h [T_{el}]^T \int_0^1 \int_0^1 \begin{bmatrix} [F_c]^T \\ -1 \end{bmatrix} \left[[F_c] - 1 \right] J(s,t) \, dt \, ds [T_{el}] \qquad [2.35]$$

The matrix related to convection is therefore:

$$[K_h] = h \int_0^1 \int_0^1 \begin{bmatrix} [F_c]^T \\ -1 \end{bmatrix} \left[[F_c] - 1 \right] J(s,t) \, dt \, ds \qquad [2.36]$$

In the case of a face with dimensions (1 m × 1 m), the convective conductivity matrix is:

$$[K_h] = \frac{h}{36} \begin{bmatrix} 4 & 2 & 1 & 2 & -9 \\ 2 & 4 & 2 & 1 & -9 \\ 1 & 2 & 4 & 2 & -9 \\ 2 & 1 & 2 & 4 & -9 \\ -9 & -9 & -9 & -9 & 36 \end{bmatrix} \qquad [2.37]$$

Like the conductivity matrix of the solid, this matrix is singular. This means that, with or without convection, at least one node (real or virtual) must be fixed in order to make the conductivity matrix positive definite. To solve a problem involving conduction and convection, we need to calculate two types of conductivity matrices. The first is called $[K_k]$ and the second $[K_h]$. The two matrices must be added together. The $[K_h]$ matrices carry the additional degrees of freedom corresponding to the virtual convective nodes:

$$K = K_k + K_h \qquad [2.38]$$

2.3.2. *Solid cube and cube with cavity subjected to convection*

Figure 2.5 deals with a cube, the domain being composed of one volume element and two convection elements. Convection is imposed on two opposite vertical faces and a constant flux totaling 240 *W* on the upper face (four nodal fluxes of 60 W). The same virtual node is used for the two convective vertical faces; its temperature is set at 300 K. The conduction coefficient k is equal to 10 W $K^{-1}m^{-1}$, and the convection coefficient h is equal to 20 W K^{-1} m^{-2}. The dissipation is 1260 *WK*, the temperature of the upper face is 310.5 K and that of the lower face is 301.5 K.

Figure 2.5. *Solid cube, one face loaded and two faces subject to convection*

In the example in Figure 2.6, we consider the case of a hollow cube. On the left, we see a wireframe representation of the box. The finite element model consists of six hexahedrons, which are represented by their edges.

In this example, we use the same boundary conditions as in the previous problem, with a single convection node. The dissipation is of the same order of magnitude, namely 1,659 *WK* instead of 1,260 *WK* for the solid cube.

The temperature of the upper face is 314 K and that of the lower face is 298 K. On the inner horizontal faces, the temperatures are 313 K and 299 K. The ratio of the dimensions between the interior and exterior is 0.8. The model includes six conduction elements and two convection elements.

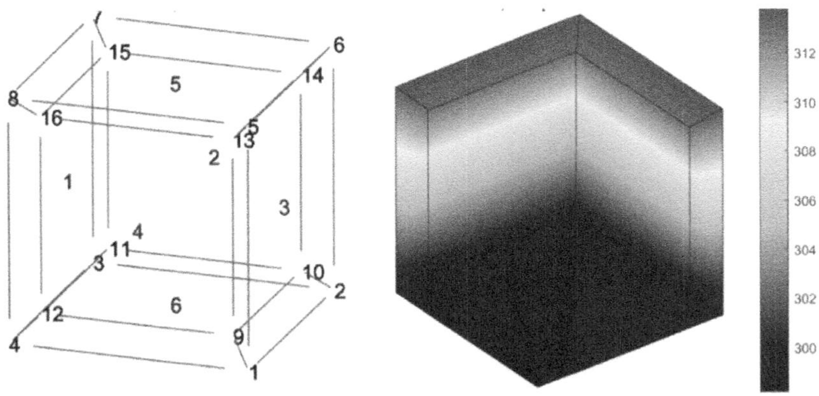

Figure 2.6. *Wireframe diagram and interior view of three of the six hexahedrons*

2.3.3. *Refinement of the container mesh*

The model based on a mesh of eight hexahedrons (Figure 2.7) gives an initial idea of the solution, but it is very imperfect. Here, the wireframe representation only concerns the outer layer. To obtain a more realistic simulation, the mesh needs to be refined, for example by imposing a mesh of 96 hexahedrons (Figure 2.8). The bottom of the figure shows the isotherms on the outer and inner skins. Two cuts were also made in the vertical planes based on the diagonals of the horizontal faces. The temperature of the virtual node on the upper face is 333.3 K. It is 335 K in the model with 16 elements per face.

The dissipation is 3995 *WK*. It increases to 4192 *WK* in the model with $16 \times 6 = 96$ elements (Figure 2.8). When examining the isotherms, slight asymmetries can be seen due to imperfections in the Matlab© interpolation method. This is because the "fill" function is based on simple linear interpolation in a triangle. In the examples in Figures 2.7 and 2.8, the cavity is considered adiabatic.

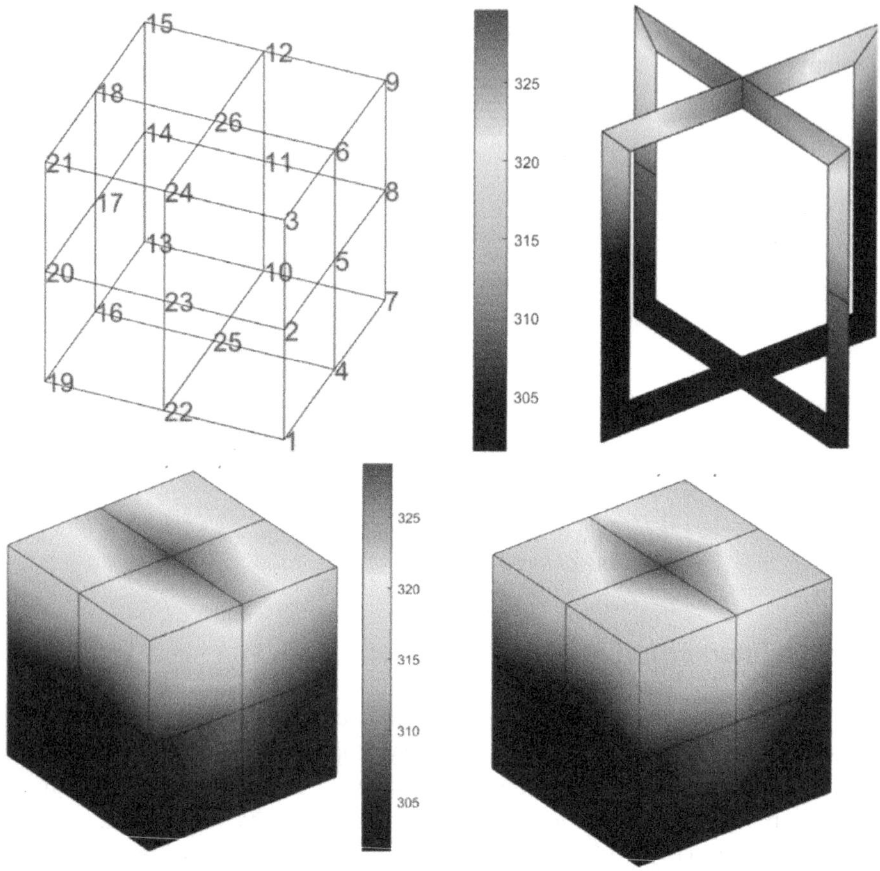

Figure 2.7. *Container comprising eight volume elements and three convective faces*

Refining the mesh in Figure 2.7 leads to the results shown in Figure 2.8. The temperatures of the outer layer of the sections in the $x = 0$ and $y = 0$ planes are shown in the lower part of the figure.

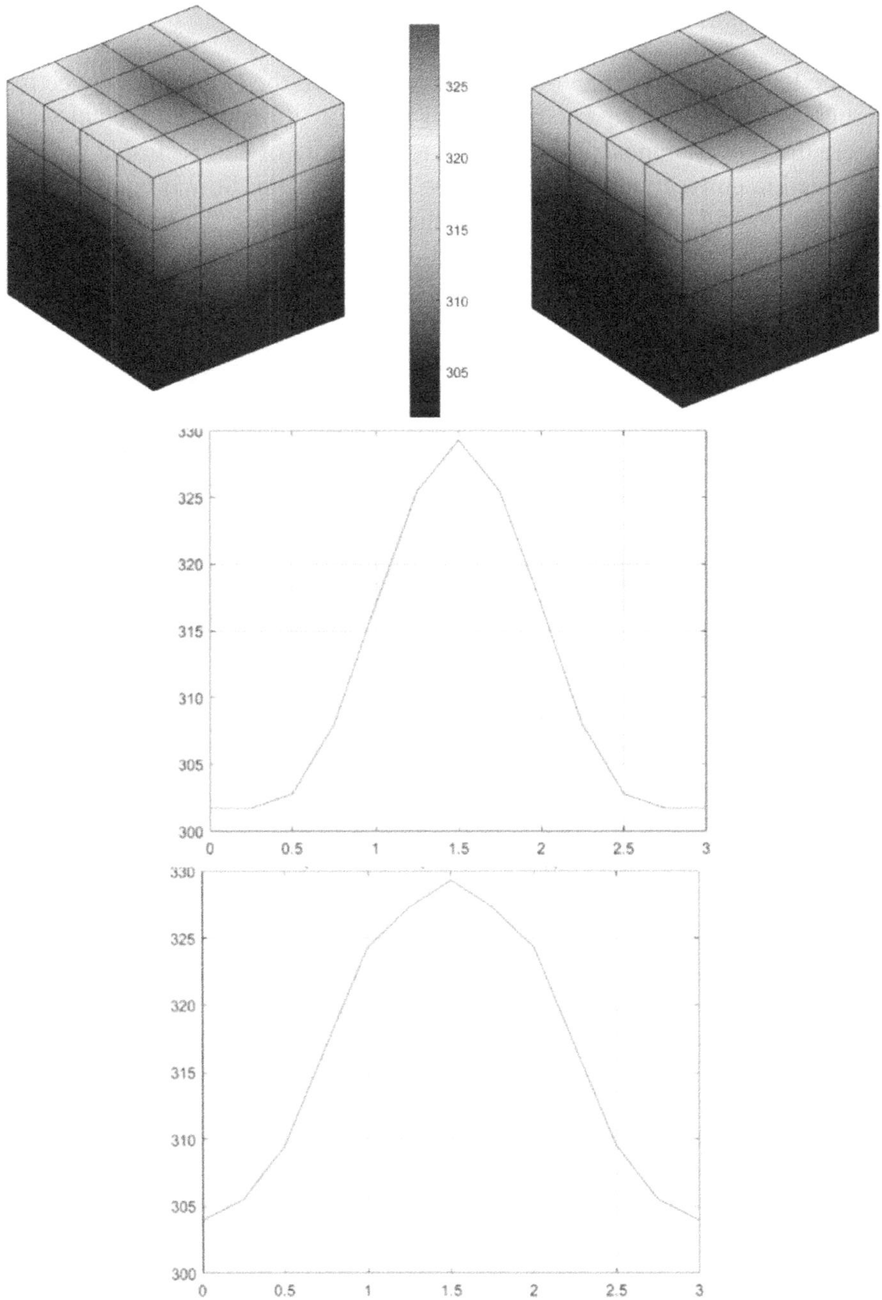

Figure 2.8. *Container comprising 96 volume elements and three convective faces*

2.3.4. *Convection within the cavity inside the cube*

The temperature of the virtual node common to the two outer convective vertical walls is 300 K. For the virtual node of the upper horizontal wall, where the flow is imposed, we obtain a temperature of 326 K. Inside the box subjected to convection, the temperature is free and the flow is equal to 0. A temperature of 308 K is obtained. At these three nodes, the heat fluxes are –240, 240 and 0 *W*. The thickness of the box is 0.244 m. The dissipation is 3130 WK (Figure 2.9).

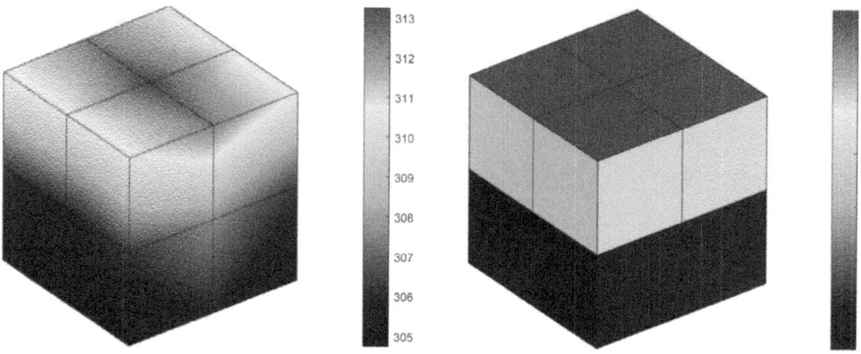

Figure 2.9. *Nodal and average temperatures in convective cavity (3130 WK)*

When switching to 16 elements per cube face, the results do not change much. The temperature of the virtual node on the top face goes up to 327 K and the dissipation to 3228 *WK*.

The temperature profiles on the contours of the $x = 0$ and $y = 0$ sections are shown in Figure 2.10. It is interesting to show the isotherms in the sections, for example, in the $x = 0$ plane (Figures 2.11–2.14).

They provide a better understanding of volume behavior. These sections show that the solution is not greatly affected by the number of layers used to model the box walls. Figure 2.14 compares the isotherms in the $x = 0$ and $y = 0$ sections.

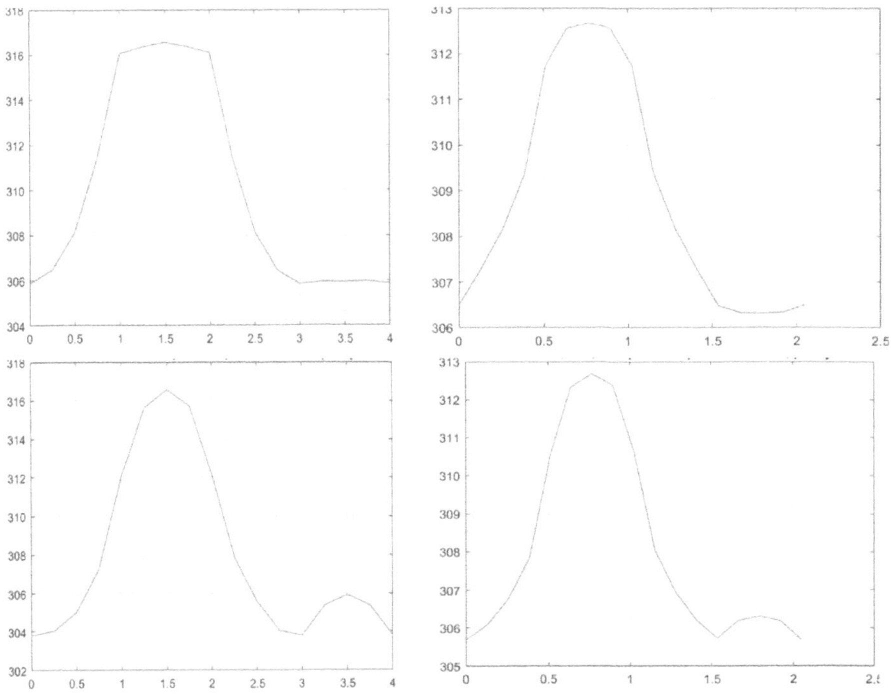

a) b)

Figure 2.10. *Sections x = 0, y = 0, exterior (a), interior (b)*

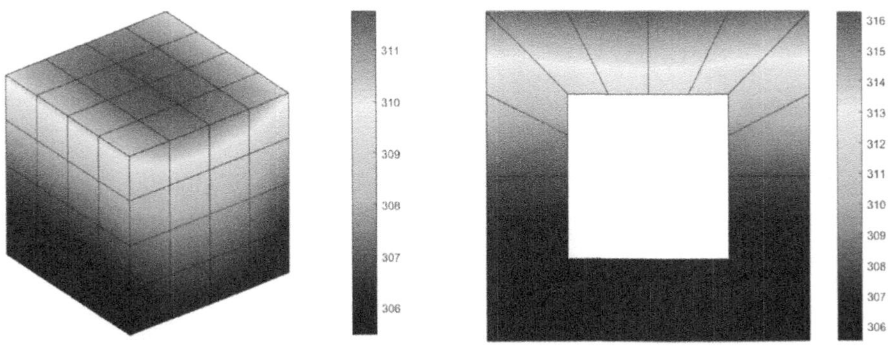

Figure 2.11. *One element on the thickness of the container*

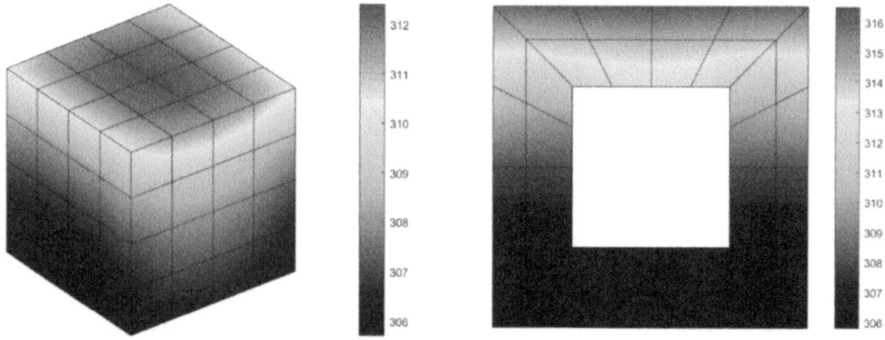

Figure 2.12. *Two elements on the thickness of the container*

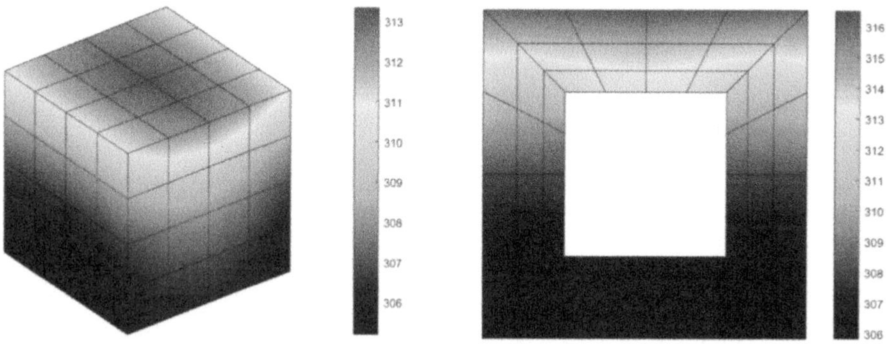

Figure 2.13. *Three elements on the thickness of the container*

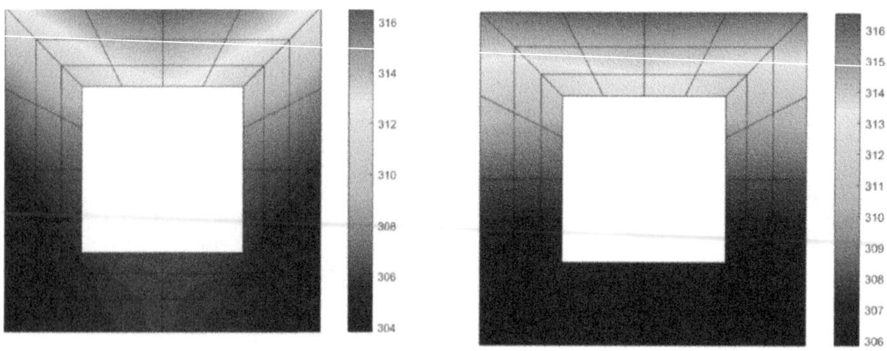

Figure 2.14. *Three elements on the thickness of the container, cuts in the plane y = 0*

2.3.5. *Adiabatic cavity*

In radiation applications involving the radiosity variable, it is useful to calculate the average temperatures of surface elements.

The results of Figure 2.9 are shown in Figure 2.15, and the results of Figure 2.13 are shown in Figure 2.16, where the average temperatures in each of the interior surface elements are used.

The temperature of the virtual node common to the two outer convective vertical walls is 300 K. The temperature of the virtual node of the upper horizontal wall is 327 K. At the two virtual nodes, the heat fluxes are –240 *W* and 240 *W*.

The thickness of the box is 0.244 m, the cavity area is 2.46 m^2 and the dissipation is 3228 *WK*. The average, minimum and maximum temperatures of the cavity are 308 K, 303 K and 318 K at iteration no. 0.

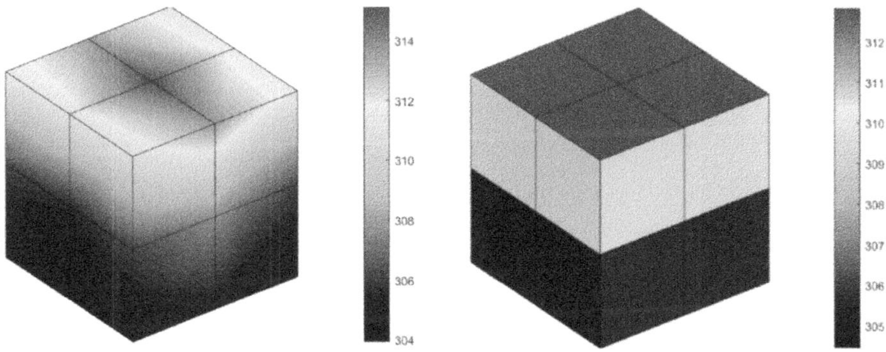

Figure 2.15. *Nodal and average temperatures in the adiabatic cavity*

When switching to 16 elements per cube face, the results change little: at iteration no. 0, the temperature of the virtual node on the upper face rises to 328 K, dissipation to 3,328 *WK* (a 3% increase), and the average, minimum and maximum temperatures to 308 K, 303 K and 318 K.

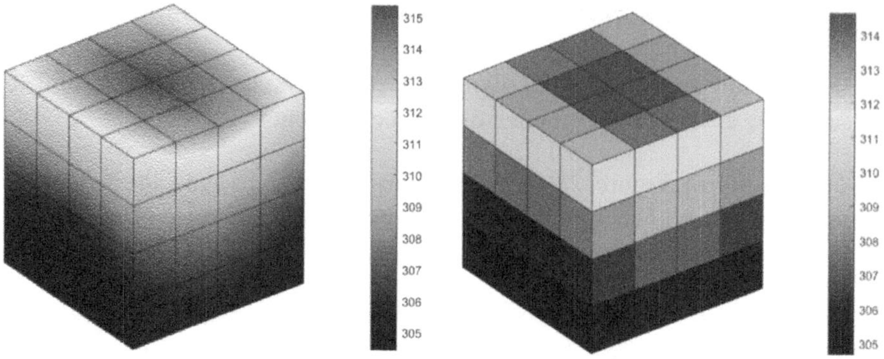

Figure 2.16. *Adiabatic cavity modeled with a fairly fine mesh*

3

Transient Response

Two new physical quantities are introduced: the density of the material and its heat capacity (Lewis et al. 2004). The specific heat capacity c_p (or specific heat), expressed in J kg^{-1} K^{-1}, corresponds to a system defined by unit of mass, $c_v = \rho\ c_p$ (the term *specific heat* is sometimes used). *Thermal capacity* (J K^{-1}) is an extensive scalar quantity; its conjugate is the temperature. The thermal diffusivity α of a material, expressed in m^2 s^{-1}, represents its tendency to facilitate heat diffusion:

$$\alpha = \frac{k}{\rho\ c_p} \quad [3.1]$$

3.1. General formulation

The differential equation for transient heat transfer locally expresses the energy conservation (Washizu 1955; Hogge 1970):

$$k\ \Delta\tau + c_v \frac{\partial \tau}{\partial t} = -\ \bar{Q}\ (x, y, z, t) \quad [3.2]$$

The variation in heat flow increased by the rate of increase in *enthalpy* measured in joules (*J*), which corresponds to the heat absorbed, is balanced by the imposed heat flow $\bar{Q}$ coming from the volume sources. The variable c_v (J m^{-3} K^{-1}) represents the heat capacity per unit volume, and the term k (W m^{-1} K^{-1}) represents the conductivity coefficient, which is assumed here to be constant.

If the boundary conditions remain unchanged during time interval Δt, such that the enthalpy rate can be considered as a function of spatial coordinates only, the equilibrium of the integral used in steady state can be modified to accept [3.4] everywhere in Ω:

$$< I(\tau) = \left[\int_{\Omega} \frac{1}{2} k \, \vec{\nabla} \tau . \vec{\nabla} \tau - \bar{Q} \, \tau + c_v \frac{\partial \tau}{\partial t} \tau \right] d\Omega + \int_{S_2} \bar{q}_n \, \tau \, dS + \frac{h}{2} \int_{S_3} \left(\tau - \tau_f \right)^2 dS > Min \tag{3.3}$$

The temperature field is expressed in terms of nodal temperatures and shape functions.

Using the Rayleigh–Ritz method, as in relations [1.26]–[1.28], the term in the previous integral comprising the time derivative of the temperature is transformed as follows:

$$\int_{\Omega} c_v \frac{\partial \tau}{\partial t} \tau \, d\Omega = [B]^T \int_{\Omega} c_v \, [W]^T [W] \, d\Omega \left[\dot{B} \right] \tag{3.4}$$

The temporal derivation of the temperature field is transferred to the Rayleigh–Ritz parameters $[B]$. The second member of the previous relation can be rewritten as:

$$[B]^T \int_{\Omega} c_v \, [W]^T [W] \, d\Omega \left[\dot{B} \right] = [B]^T [C] \left[\dot{B} \right] \tag{3.5}$$

This relationship includes a new entity: the *capacity matrix* $[C]$:

$$[C] = \int_{\Omega} c_v \, [W]^T [W] \, d\Omega \tag{3.6}$$

The matrix $[C]$ has dimensions equal to the product of the specific heat c_v (J m^{-3} K^{-1}) and the volume, that is, J K^{-1}. If we replace the Rayleigh–Ritz parameters with those of the finite elements, the volume loads and the Neumann boundary condition loads of [3.3] become:

$$\int_{\Omega} -\bar{Q}\tau \, d\Omega + \int_{S_2} \bar{q}_n \; \tau \, dS \;= -[T]^T [f] \; Min \qquad [3.7]$$

Moving on to finite elements, expression [3.5], in which to condense the formulation the term related to convection has not been included, becomes:

$$I = \frac{1}{2}[T]^T [K][T] + [T]^T [C]\left[\dot{T}\right] - [T]^T [f] \; Min \qquad [3.8]$$

From this relationship, we can deduce a matrix formulation by writing Euler's equations for the stationarity conditions of [3.8]:

$$[K][T] + [C]\left[\dot{T}\right] = [f] \qquad [3.9]$$

3.2. Temporal integration

The essential assumption of the step-by-step integration method postulates a linear temporal variation of the temperature field, such that:

$$\frac{[T(t)] - [T(t-\Delta t)]}{\Delta t} = \frac{1}{2}\left(\left[\dot{T}(t)\right] + \left[\dot{T}(t-\Delta t)\right]\right) \qquad [3.10]$$

From this, we can deduce the value of the time derivative of the temperature at time t:

$$\left[\dot{T}(t)\right] = \frac{2}{\Delta t}\{[T(t)] - [T(t-\Delta t)]\} - \left[\dot{T}(t-\Delta t)\right] \qquad [3.11]$$

$$\left\{[K] + \frac{2}{\Delta t}[C]\right\}[T(t)] = \\ [f(t)] + [C]\left\{\frac{2}{\Delta t}[T(t-\Delta t)] + \left[\dot{T}(t-\Delta t)\right]\right\} \qquad [3.12]$$

We approximate the temporal derivative of temperature using a finite difference over:

$$\left[\dot{T}(t-\Delta t)\right]=\frac{[T(t)]-[T(t-\Delta t)]}{\Delta t} \qquad [3.13]$$

It allows us to know the nodal temperatures at time t if we know them, as well as their derivatives at time $t-\Delta t$. All we need to do is construct the matrices K and C and update the boundary conditions to calculate the nodal temperature at each time t. This equation can be rewritten using the formalism of an iterative calculation process, where we replace $T(t)$ with T^{n+1}, $T(t-\Delta t)$ with T^{n}, and $f(t)$ with f^{n+1}:

$$\{[C]+\Delta t[K]\}\left[T^{n+1}\right]=[C]\left[T^{n}\right]+\Delta t\left[f^{n+1}\right] \qquad [3.14]$$

This expression corresponds to what is known as the *implicit scheme.*

The column matrix $[T]$ can be divided into two parts:

– unknown temperatures $[T_l]$;

– fixed and therefore constant temperatures $[T_f]$. They do not depend on iteration.

We therefore rewrite:

$$\begin{gathered}\left(\begin{bmatrix}C_{11} & C_{1f}\end{bmatrix}+\Delta t\begin{bmatrix}K_{11} & K_{1f}\end{bmatrix}\right)\begin{bmatrix}T^{n+1}\\ T_f\end{bmatrix}=\\ \begin{bmatrix}C_{11} & C_{1f}\end{bmatrix}\begin{bmatrix}T^{n}\\ T_f\end{bmatrix}+\Delta t\begin{bmatrix}f^{n+1}\end{bmatrix}\end{gathered} \qquad [3.15]$$

Expanding on this:

$$\begin{gathered}\left([C_{11}]+\Delta t\,[K_{11}]\right)\left[T^{n+1}\right]=\\ [C_{11}]\left[T^{n}\right]+\Delta t\left(\left[f^{n+1}\right]-\left[K_{1f}\right]\left[T_f\right]\right)\end{gathered} \qquad [3.16]$$

$$\left[C_{11}\right]\left[T_1^{n+1}\right] + \Delta t \left[K_{11} \quad K_{1f}\right]\begin{bmatrix} T_1^{n+1} \\ T_f \end{bmatrix} = \left[C_{11}\right]\left[T_1^n\right] + \Delta t \left[f^{n+1}\right] \tag{3.17}$$

$$\left[T^{n+1}\right] = \left(\left[C_{11}\right] + \Delta t\left[K_{11}\right]\right)^{-1} \begin{Bmatrix} \left[C_{11}\right]\left[T^n\right] + \\ \Delta t\left(\left[f^{n+1}\right] - \left[K_{1f}\right]\left[T_f\right]\right) \end{Bmatrix} \tag{3.18}$$

If there is no imposed flow, we have:

$$\left[T^{n+1}\right] = \left(\left[C_{11}\right] + \Delta t\left[K_{11}\right]\right)^{-1}\left\{\left[C_{11}\right]\left[T^n\right] - \Delta t\left[K_{1f}\right]\left[T_f\right]\right\} \tag{3.19}$$

Without an imposed temperature, we have:

$$\left[T^{n+1}\right] = \left(\left[C_{11}\right] + \Delta t \left[K_{11}\right]\right)^{-1}\left\{\left[C_{11}\right]\left[T^n\right] + \Delta t\left[f^{n+1}\right]\right\} \tag{3.20}$$

If there is no imposed flow or fixation, we can remove the indices and obtain:

$$\left[T^{n+1}\right] = \left(\left[C\right] + \Delta t \left[K\right]\right)^{-1}\left[C\right]\left[T^n\right] \tag{3.21}$$

3.3. The capacity matrix

The capacity matrix [C] defined in [3.6] is a function of the density ρ of the material, its heat capacity $c_v = \rho\ c_p$, its volume V and its geometry. At the finite element level, the capacity matrix is defined from the shape functions f_i, terms of the column matrix $[F]$ of the temperature field [3.23]:

$$\left[C\right] = \int_{\Omega} \rho c_p \left[F\right]^T \left[F\right] d\Omega \tag{3.22}$$

$$\tau = \left[F\right]^T \left[T\right] \tag{3.23}$$

$$\left[F(x,y)\right]^T = [(1-x)(1-y)(1-z) + x(1-y)(1-z) + xy(1-z) + (1-x)y(1-z) + (1-x)(1-y)z + x(1-y)z + xyz + (1-x)yz] \quad [3.24]$$

$$[T]^T = [T_1 \quad T_2 \quad T_3 \quad T_4 \quad T_5 \quad T_6 \quad T_7 \quad T_8] \quad [3.25]$$

For a simple cube element, we verify that the sum of the terms in matrix [*C*] [3.26] is equal to $\rho\ c_p\ V$, which corresponds to the heat capacity of the finite element analyzed if it is concentrated at a single point. Matrix [*C*] is expressed in J K^{-1}:

$$[C] = \frac{\rho c_p V}{216}\begin{bmatrix} 8 & 4 & 2 & 4 & 4 & 2 & 1 & 2 \\ 4 & 8 & 4 & 2 & 2 & 4 & 2 & 1 \\ 2 & 4 & 8 & 4 & 1 & 2 & 4 & 2 \\ 4 & 2 & 4 & 8 & 2 & 1 & 2 & 4 \\ 4 & 2 & 1 & 2 & 8 & 4 & 2 & 4 \\ 2 & 4 & 2 & 1 & 4 & 8 & 4 & 2 \\ 1 & 2 & 4 & 2 & 2 & 4 & 8 & 4 \\ 2 & 1 & 2 & 4 & 4 & 2 & 4 & 8 \end{bmatrix} \quad [3.26]$$

3.4. Temperature evolution in a solid with adiabatic boundaries

A uniform temperature test is performed on a domain measuring 2 m × 1 m × 0.5 m with adiabatic walls. Half of the area is at 300 K, and the other half at 290 K. The temporal evolution and the moment at which the temperature becomes uniform are examined. The homogenization process depends on diffusivity [3.1].

The data for the problem are conductivity coefficient = 2 W $m^{-1}K^{-1}$, specific capacity = 1,000 J kg^{-1} K^{-1} and specific mass = 2,500 kg m^{-3}. In this example, the thermal capacity of the domain is equal to 2.5×10^6 J K^{-1}.

After 33.2 h, the temperature difference is reduced by half: 5 K. Temperature homogeneity is achieved after 180 h, when the temperature difference is equal to 0.1 K (Figure 3.1). At time intervals of 24, 48, 72 and 96 h, the temperature difference evolves as follows: 6.5 K, 3.3 K, 1.7 K and finally 0.8 K (isotherms visible in Figure 3.2).

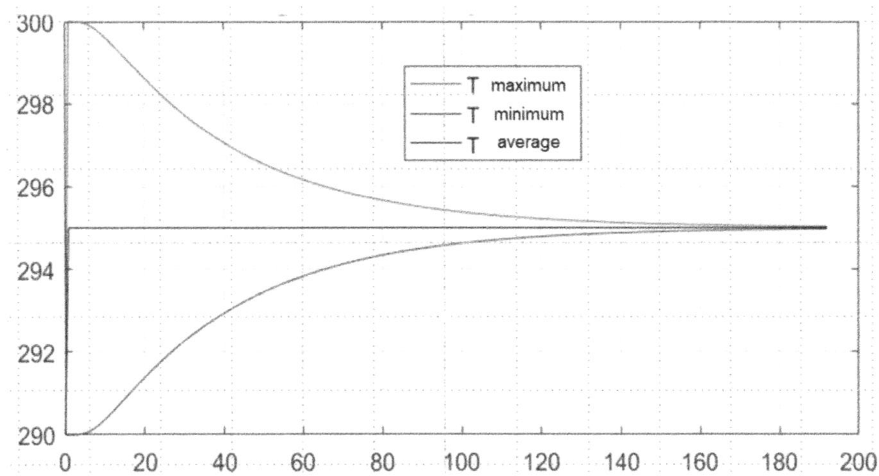

Figure 3.1. *Changes in maximum, minimum and average temperatures; 192 iterations. For a color version of this figure, see www.iste.co.uk/beckers/finite.zip*

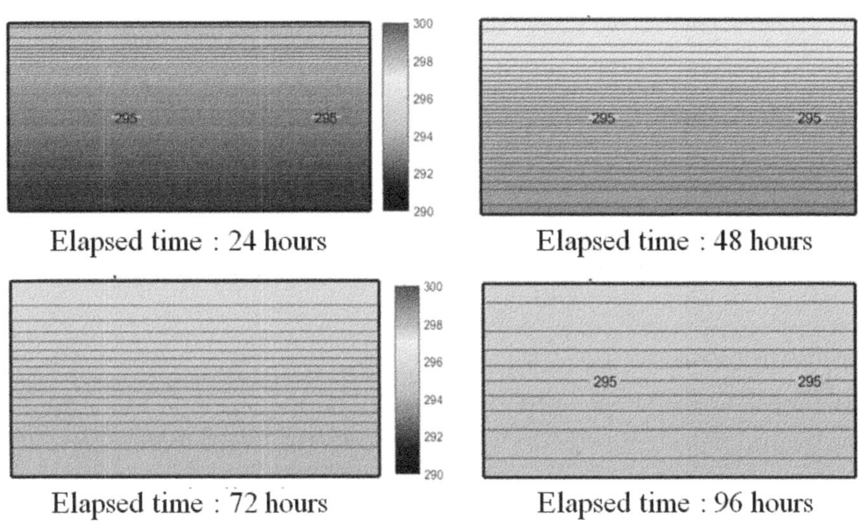

Figure 3.2. *Isotherms in a rectangular parallelepiped, 96 iterations. For a color version of this figure, see www.iste.co.uk/beckers/finite.zip*

3.5. Solid immersed in a fluid

Consider a plate at a temperature of 290 K immersed in a fluid at 300 K, the temperature of the virtual node (Figure 3.3). In a two-dimensional analysis, the two large faces of the solid are considered adiabatic. The dimensions of the solid are as follows: 1 m × 2 m × 0.1 m. The conduction coefficient is equal to 1 W K^{-1} m^{-1}, the convection coefficient to 25 W $K^{-1}m^{-2}$, the specific mass to 2,500 kg m^{-3} and the heat capacity to 1,000 J K^{-1} kg^{-1}. The capacity of the domain is 0.5 MJ K^{-1}.

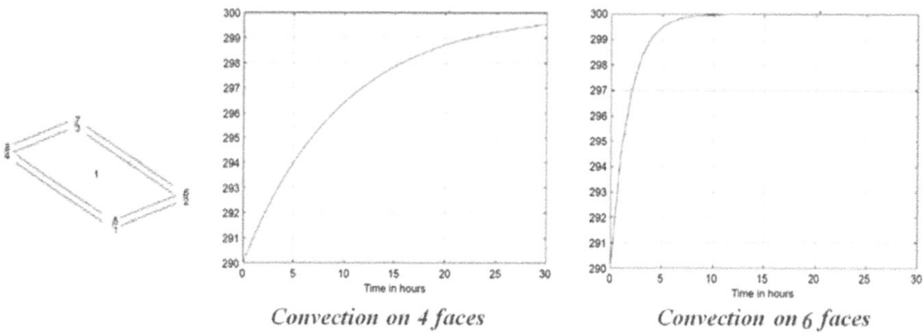

Figure 3.3. *Temporal evolution of the lowest temperature of the solid*

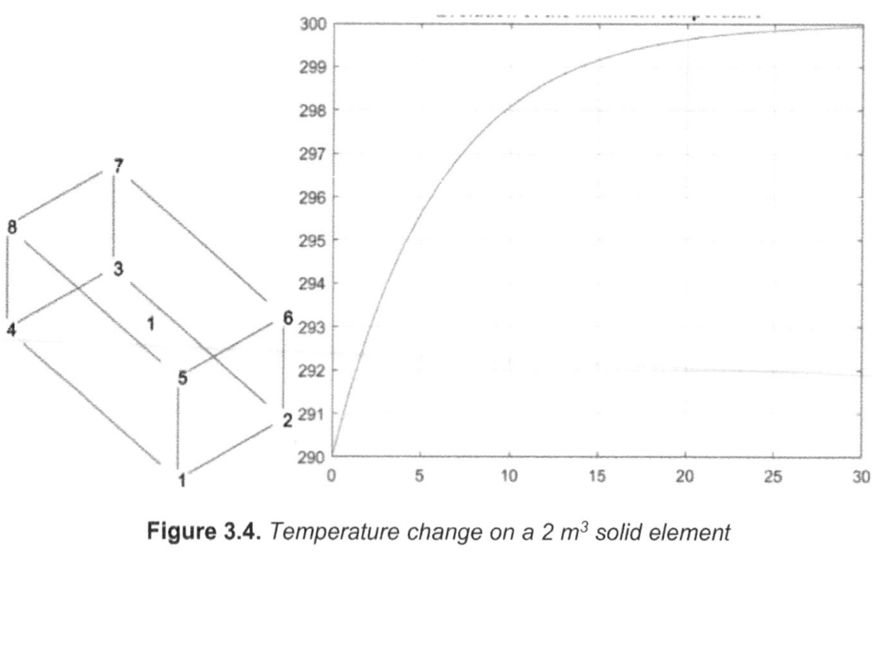

Figure 3.4. *Temperature change on a 2 m^3 solid element*

The time diagram in Figure 3.4 shows time expressed in hours on the x-axis and temperature expressed in degrees Kelvin on the y-axis. With four convective faces, the plate heats up in just over 30 h. However, with the six convective faces of the parallelepiped, the plate heats up in about 10 h.

For a massive domain measuring 2 m^3 and therefore with a total capacity of 5 MJ K^{-1}, for which the exchange takes place on six sides, the ingot is heated in 30 h (Figure 3.4).

In this problem, the eight nodes of the volume element all play the same role. Consequently, their temperatures are identical, so that only one problem with two degrees of freedom needs to be solved: the temperature of the solid and that of the virtual node. The temperature of the solid is initially 290 K and that of the virtual node is constant and set at 300 K.

The conductivity matrix of the element with eight identical nodes is zero, while that relating to convection is, for a convection coefficient of 25 W K^{-1} m^{-2}:

$$[Kc] = \begin{bmatrix} 250 & -250 \\ -250 & 250 \end{bmatrix} \qquad [3.27]$$

The capacity matrix depends on a single scalar, equal to the product of volume, specific mass, and heat capacity, that is, 5 MJ K^{-1}.

The capacity matrix $[Ca]$ is written in terms of the two degrees of freedom corresponding to the temperatures of the solid and the virtual node. It is expressed in J K^{-1}:

$$[Ca] = \begin{bmatrix} 5{,}000{,}000 & 0 \\ 0 & 0 \end{bmatrix} \qquad [3.28]$$

In the following calculations, the time step is 1 h, that is, $\Delta t = 3{,}600$ s. The temperature T^{n+1} at iteration n+1 is calculated using equation [3.28], in which the load $[f^{n+1}]$ is zero, the temperature $Tf = 300$ K and $T\,n = 290$ K:

$$\begin{gathered} \left[T^{n+1}\right] = \left(\left[Ca(1,1)\right] + \Delta t\ \left[Kc(1,1)\right]\right)^{-1} \\ \left\{\left[Ca(1,1)\right]\left[T^{n}\right] - \Delta t\left(\ Kc(1,2)\left[T_f\right]\ \right)\right\} \end{gathered} \qquad [3.29]$$

The result is: $[T^{n+1}] = 291.5$ K.

This way of solving the problem is similar to a *nodal method*, in which geometry is absent.

4

Isoparametric Elements

Steven Anson Coons' (1967, 1977) ideas concerning meshes defining geometries were extended to finite elements by Bruce Irons (Irons 1966; Ergatoudis et al. 1968). In this chapter, they are explained in the case of volume elements.

4.1. The concept of Coons patch

The best-known application of Coons' method is the technique known as *lofting* (Beckers 2017). It consists of generating a surface by linearly evolving a profile from an initial shape (composition of several lines in red on the left) to another shape (arc of a circle, in red on the right) (Figure 4.1). This is a widely used technique for defining aircraft wing profiles.

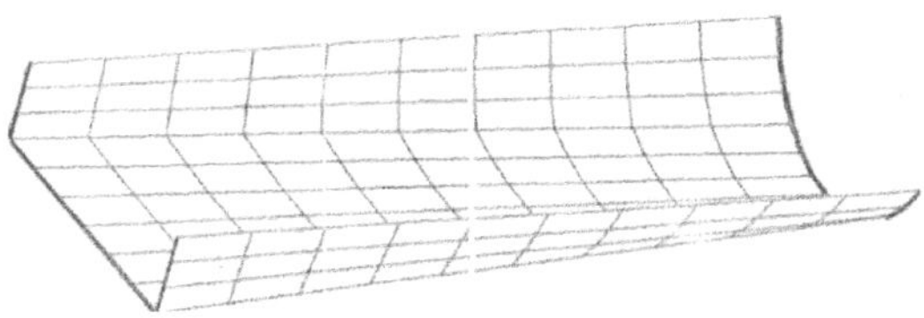

Figure 4.1. *Lofting technique*

The simplest interpolation situation is to move from any line to any other line (Figure 4.2).

For a color version of all the figures in this chapter, see www.iste.co.uk/beckers/finite.zip.

The four sides of the classic bilinear *Coons patch* are line segments. In this example, starting with a quadrilateral, which is a polygon (and therefore a flat object), we simply added a slight twist to transform it into a hyperbolic paraboloid.

A Coons patch can be easily represented by drawing the networks of lines based on the two pairs of opposite sides (the black and red ones, Figure 4.2). With a large twist, we obtain the result shown in Figure 4.3.

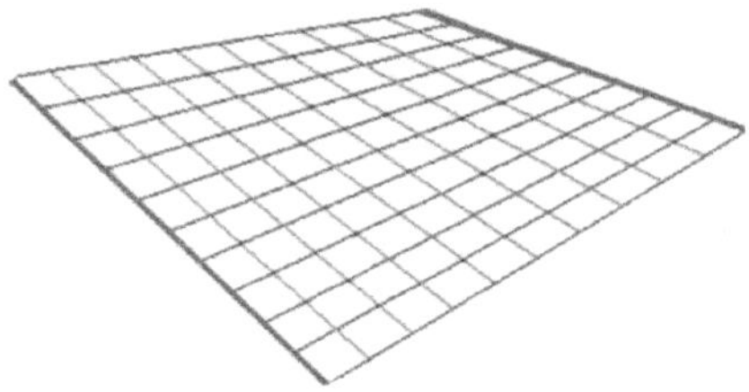

Figure 4.2. *Lofting between two line segments*

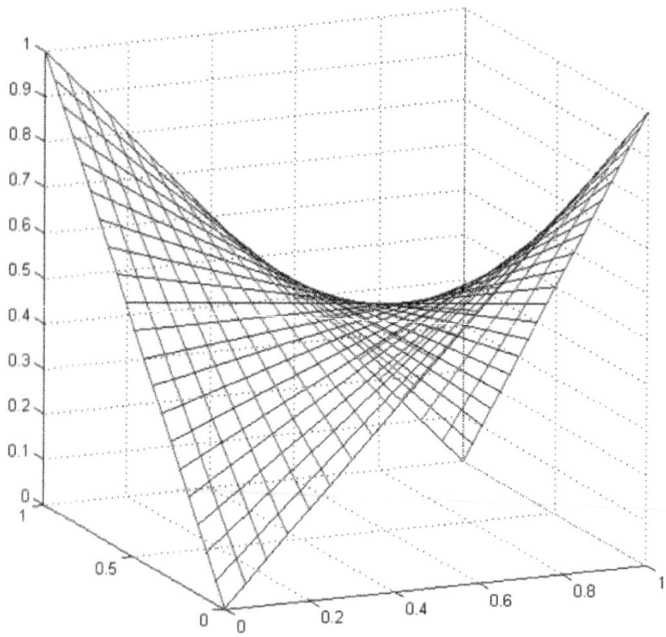

Figure 4.3. *Gridding of the bilinear Coons patch*

4.2. Trilinear hexahedron

A *trilinear hexahedron* is a hexahedron whose 12 edges are line segments. This volume is defined by eight vertices:

$$[Q]=\begin{bmatrix} Q_1 \\ Q_2 \\ \dots \\ Q_8 \end{bmatrix}=\begin{bmatrix} x_1 & y_1 & z_1 \\ x_2 & \dots & \dots \\ \dots & \dots & \dots \\ x_8 & \dots & z_8 \end{bmatrix} \quad [4.1]$$

$[Q]$ is an eight-row, three-column matrix containing the Cartesian coordinates of these eight vertices. The shape functions of the eight points depend on the three parametric variables u, v and w. They are equal to 1 on one of the vertices and 0 on the others. They are stored in the row matrix of shape functions $[F]$:

$$P(u,v,w)^T=[x(u,v,w)\ y(u,v,w)\ z(u,v,w)]=[F][Q] \quad [4.2]$$

The points of the hexahedron are defined by their three parametric coordinates u, v and w, which vary between 0 and 1, the eight vertices defined in [4.1], and the function F, defined in [4.3], but depending on the parameters x and y:

$$[F] = \begin{bmatrix} f_1 & f_2 & f_3 & f_4 & f_5 & f_6 & f_7 & f_8 \end{bmatrix}^T =$$

$$\begin{bmatrix} u(1-v)(1-w) & (1-u)(1-v)w \\ u(1-v)(1-w) & u(1-v)w \\ uv(1-w) & uvw \\ (1-u)v(1-w) & (1-u)vw \end{bmatrix}^T \quad [4.3]$$

All functions f_i satisfy condition [2.23]:

$$f_1+f_2+f_3+f_4+f_5+f_6+f_7+f_8=1 \quad [4.4]$$

The column matrices [*X*], [*Y*] and [*Z*] contain the *x*, *y* and *z* coordinates of the eight vertices. Expression [4.1] can then be written as:

$$[Q]=\begin{bmatrix} Q_1 \\ Q_2 \\ \dots \\ Q_8 \end{bmatrix}=\begin{bmatrix}[X] & [Y] & [Z]\end{bmatrix} \qquad [4.5]$$

Relation [4.2] becomes:

$$\begin{aligned} x(u,v,w) &= [F][X] \\ y(u,v,w) &= [F][Y] \\ z(u,v,w) &= [F][Z] \end{aligned} \qquad [4.6]$$

The Jacobian matrix of the element is written as:

$$[J]=\begin{bmatrix} \frac{\partial x}{\partial u} & \frac{\partial y}{\partial u} & \frac{\partial z}{\partial u} \\ \frac{\partial x}{\partial v} & \frac{\partial y}{\partial v} & \frac{\partial z}{\partial v} \\ \frac{\partial x}{\partial w} & \frac{\partial y}{\partial w} & \frac{\partial z}{\partial w} \end{bmatrix} \qquad [4.7]$$

To transform an integration in Cartesian space into an equivalent integration in parametric space, we use the Jacobian *J*, which is the determinant of the Jacobian matrix [*J (u, v, w)*]:

$$\boldsymbol{J}=\det\left(\left[J(u,v,w)\right]\right)=\frac{\partial x}{\partial u}\left(\frac{\partial y}{\partial v}\frac{\partial z}{\partial w}-\frac{\partial y}{\partial w}\frac{\partial z}{\partial v}\right)- \frac{\partial y}{\partial u}\left(\frac{\partial x}{\partial v}\frac{\partial z}{\partial w}-\frac{\partial x}{\partial w}\frac{\partial z}{\partial v}\right)+\frac{\partial z}{\partial u}\left(\frac{\partial x}{\partial v}\frac{\partial y}{\partial w}-\frac{\partial x}{\partial w}\frac{\partial y}{\partial v}\right) \qquad [4.8]$$

The volume is calculated either in Cartesian coordinates *x*, *y* and *z* or in intrinsic coordinates *u*, *v* and *w*:

$$\iiint_{\Omega} dxdydz=\int_0^1\int_0^1\int_0^1 \boldsymbol{J}(u,v,w)\; du\; dv\; dw \qquad [4.9]$$

To calculate the integral of a function $f(x, y)$ over the volume, we use *Gaussian quadrature*. The technique consists of evaluating the integrand $f(x, y, z)$ at a certain number of points. We then simply calculate the sum of its values, weighted by assigning a weight W_i to each of the evaluation points:

$$\iiint_{\Omega} f(x,y,z)\ dxdydz = \sum f(x_i,y_i,z_i)\ W_i \qquad [4.10]$$

For a parametric function, the Jacobian transformation must be added. The function to be integrated must be expressed in parametric coordinates:

$$\begin{aligned}\iiint_{\Omega} f(x,y,z)\ dxdydz &= \int_0^1\int_0^1\int_0^1 g(u,v,w)\boldsymbol{J}(u,v,w)\ dudvdw \\ &= \sum_i g(u_i,v_i,t_i)W_i\boldsymbol{J}(u_i,v_i,w_i)\end{aligned} \qquad [4.11]$$

4.3. Two-parameter surfaces or Coons patches

The patch is defined in three-dimensional space by its four vertices:

$$[Q] = \begin{bmatrix} Q_1 \\ Q_2 \\ Q_3 \\ Q_4 \end{bmatrix} = \begin{bmatrix} x_1 & y_1 & z_1 \\ x_2 & \cdots & \cdots \\ x_3 & \cdots & \cdots \\ x_4 & \cdots & z_4 \end{bmatrix} \qquad [4.12]$$

$[Q]$ is a matrix with four rows and three columns containing the Cartesian coordinates of the four vertices. The shape functions of the four points depend on the parametric variables s and t [4.13] and are equal to 1 on one of the vertices and 0 on the others. They are stored in the row matrix $[F]$:

$$\begin{aligned}&[F_S] = [f_1 \quad f_2 \quad f_3 \quad f_4] = \\ &[(1-s)(1-t) \quad s(1-t) \quad st \quad (1-s)t]\end{aligned} \qquad [4.13]$$

The points of the patch are therefore defined through the shape functions by their two parametric coordinates s and t:

$$P(s,t)^T = [x(s,t) \quad y(s,t) \quad z(s,t)] = [F_S][Q] \qquad [4.14]$$

The functions f_i verify condition [2.23]:

$$f_1 + f_2 + f_3 + f_4 = 1 \qquad [4.15]$$

The column matrices $[X]$, $[Y]$ and $[Z]$ contain the x, y and z coordinates of the four vertices. Equation [4.12] can then be written as:

$$[Q] = \begin{bmatrix} Q_1 \\ Q_2 \\ Q_3 \\ Q_4 \end{bmatrix} = \begin{bmatrix} [X] & [Y] & [Z] \end{bmatrix} \qquad [4.16]$$

This equation becomes:

$$\begin{aligned} x(s,t) &= [F_S][X] \\ y(s,t) &= [F_S][Y] \\ z(s,t) &= [F_S][Z] \end{aligned} \qquad [4.17]$$

If the bilinear element is defined in two-dimensional space, the last line of [4.17] must be deleted.

In *2D*, the Jacobian matrix of the element is written as:

$$\begin{aligned} [J] &= \begin{bmatrix} \frac{\partial x}{\partial s} & \frac{\partial y}{\partial s} \\ \frac{\partial x}{\partial t} & \frac{\partial y}{\partial t} \end{bmatrix} \\ \frac{\partial x}{\partial s} &= \begin{bmatrix} -(1-t) & (1-t) & t & -t \end{bmatrix}[X] \\ \frac{\partial y}{\partial s} &= \begin{bmatrix} -(1-t) & (1-t) & t & -t \end{bmatrix}[Y] \\ \frac{\partial x}{\partial t} &= \begin{bmatrix} -(1-s) & -s & s & (1-s) \end{bmatrix}[X] \\ \frac{\partial y}{\partial t} &= \begin{bmatrix} -(1-s) & -s & s & (1-s) \end{bmatrix}[Y] \end{aligned} \qquad [4.18]$$

To transform an integration in Cartesian space into an equivalent integration in parametric space, we use the Jacobian *J*, which is the determinant of the Jacobian matrix [*J (s, t)*]:

$$\boldsymbol{J} = \det\left(\left[J(s,t)\right]\right) = \frac{\partial x}{\partial s}\frac{\partial y}{\partial t} - \frac{\partial x}{\partial t}\frac{\partial y}{\partial s} \qquad [4.19]$$

The area of the patch is calculated using the following formula, expressed either in Cartesian coordinates *x*, *y*, or in intrinsic coordinates *s*, *t*:

$$\iint_{\Omega_{Coons}} dxdy = \int_0^1\int_0^1 J(s,t)\ dsdt \qquad [4.20]$$

The integration of functions on surface entities is performed using Gaussian quadrature. The technique consists of evaluating the integral of the function *f* (*x*, *y*) at a certain number of points. It is then sufficient to calculate the sum of its weighted values by assigning *a priori* weight w_i to each of the evaluation points:

$$\iint_{Coons} f(x,y)dxdy = \sum f(x_i, y_i)w_i \qquad [4.21]$$

Relationships between Cartesian coordinates and parametric coordinates	
Cartesian space **Coordinates: *x, y*** ***dS* = *dx dy***	**Parametric space** **Coordinates: *s, t*** ***dS* = ?**
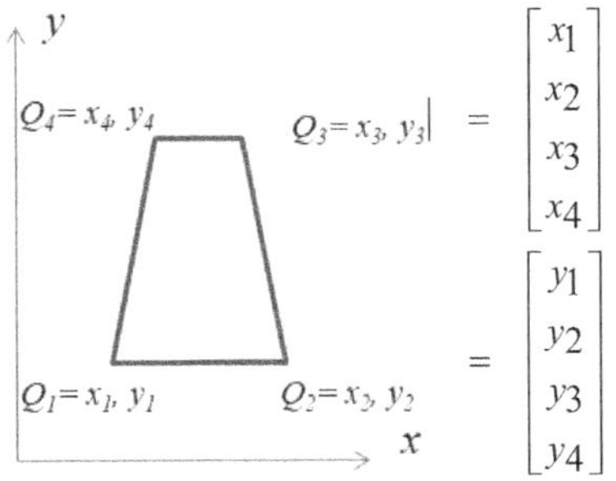	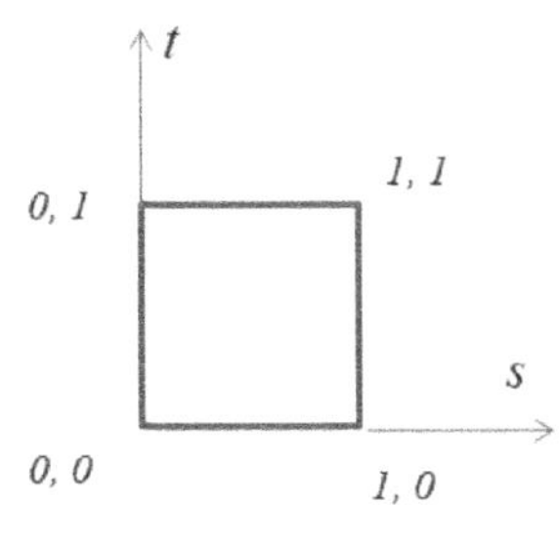

Table 4.1. *Two coordinate systems*

When the function is written in parametric coordinates, the Jacobian transformation must be added:

$$\iint_{Coons} g^{*}(x,y)dxdy = \int_0^1\int_0^1 g(s,t)\boldsymbol{J}(s,t)\ dsdt = \sum_i g(s_i,t_i)w_i\boldsymbol{J}(s_i,t_i) \qquad [4.22]$$

Gaussian quadrature in dimension 1		
Number	**Position in the interval [0, 1]**	**Weight w_i**
1	0.5	1
2	$0.5 \pm \sqrt{3}/6$	5
3	$0.5,\ 0.5 \pm \sqrt{3/20}$	4/9, 5/18, 5/18
4	$0.5 \pm 1/70\sqrt{525-70\sqrt{30}}$ $0.5 \pm 1/70\sqrt{525+70\sqrt{30}}$	$1/4+\sqrt{30}/72\ ;\ 1/4+\sqrt{30}/72$ $1/4-\sqrt{30}/72\ ;\ 1/4-\sqrt{30}/72$

Table 4.2. *Different levels of Gaussian quadrature*

The results in Table 4.3 confirm that $n \times n$ integration points are necessary to obtain an exact solution for the integration of polynomials of degree $2n$-1.

Evaluated function	**Exact solution**	**Number of Gauss points**			
		1	**4 = 2 × 2**	**9 = 3 × 3**	**16 = 4 × 4**
sin (πs)	0.6366	1	0.6162	0.6371	0.6366
s^3	0.25	0.125	0.25	0.25	0.25
s^5	0.1667	0.0313	0.1528	0.1667	0.1667
s^7	0.1250	0.0078	0.0949	0.1238	0.1250

Table 4.3. *Gaussian quadratures on a unit square*

4.4. Three-dimensional isoparametric element

An element is called *isoparametric* if it has the same shape functions for the geometry and for the field being studied, in this case temperature τ. In two dimensions, this property is written as:

$$\begin{aligned} \tau(s,t) &= [F][T] \\ x(s,t) &= [F][X] \\ y(s,t) &= [F][Y] \end{aligned} \qquad [4.23]$$

In this expression, $[T]$ represents the nodal temperatures, $[X]$ the x-coordinates of the nodes and $[Y]$ their y-coordinates. The matrix $[F]$ of shape functions is the same for all three fields.

The gradient of a scalar, for example, temperature τ (s, t), is easily calculated in parametric coordinates, but it must be expressed in Cartesian coordinates that represent the real world:

$$\begin{bmatrix} \dfrac{\partial \tau}{\partial s} \\ \dfrac{\partial \tau}{\partial t} \end{bmatrix} = \begin{bmatrix} \dfrac{\partial x}{\partial s} & \dfrac{\partial y}{\partial s} \\ \dfrac{\partial x}{\partial t} & \dfrac{\partial y}{\partial t} \end{bmatrix} \begin{bmatrix} \dfrac{\partial \tau}{\partial x} \\ \dfrac{\partial \tau}{\partial y} \end{bmatrix} = [J] \begin{bmatrix} \dfrac{\partial \tau}{\partial x} \\ \dfrac{\partial \tau}{\partial y} \end{bmatrix} \qquad [4.24]$$

We therefore have:

$$\begin{bmatrix} \dfrac{\partial \tau}{\partial x} \\ \dfrac{\partial \tau}{\partial y} \end{bmatrix} = [J]^{-1} \begin{bmatrix} \dfrac{\partial \tau}{\partial s} \\ \dfrac{\partial \tau}{\partial t} \end{bmatrix} \qquad [4.25]$$

The inverse $[J]^{-1}$ of the Jacobian matrix must be known at every point in the integration domain. As with the coordinates, the temperature field is expressed in parametric coordinates as a function of the nodal values:

$$\tau = \begin{bmatrix} (1-s)(1-t) & s(1-t) & st & (1-s)t \end{bmatrix} [T] = [F][T] \qquad [4.26]$$

It is easy to calculate the gradient of τ in parametric coordinates:

$$\begin{bmatrix} \dfrac{\partial \tau}{\partial s} \\ \dfrac{\partial \tau}{\partial t} \end{bmatrix} = \begin{bmatrix} \dfrac{\partial F}{\partial s} \\ \dfrac{\partial F}{\partial t} \end{bmatrix} [T] = \begin{bmatrix} -(1-t) & (1-t) & t & -t \\ -(1-s) & -s & s & (1-s) \end{bmatrix} [T] \qquad [4.27]$$

According to [4.27]:

$$\begin{bmatrix} \dfrac{\partial \tau}{\partial x} \\ \dfrac{\partial \tau}{\partial y} \end{bmatrix} = [J]^{-1} \begin{bmatrix} \dfrac{\partial \tau}{\partial s} \\ \dfrac{\partial \tau}{\partial t} \end{bmatrix} = [J]^{-1} \begin{bmatrix} \dfrac{\partial F}{\partial s} \\ \dfrac{\partial F}{\partial t} \end{bmatrix} [T] = \qquad [4.28]$$
$$[J]^{-1} \begin{bmatrix} -(1-t) & (1-t) & t & -t \\ -(1-s) & -s & s & (1-s) \end{bmatrix} [T]$$

When the patch is located in three-dimensional space, the Jacobian calculation must be modified as follows. Equation [4.26] defining the Jacobian matrix is transformed into:

$$[J] = \begin{bmatrix} \dfrac{\partial P}{\partial s} \times \dfrac{\partial P}{\partial t} \end{bmatrix} = \begin{bmatrix} \dfrac{\partial y}{\partial s}\dfrac{\partial z}{\partial t} - \dfrac{\partial y}{\partial t}\dfrac{\partial z}{\partial s} & \dfrac{\partial z}{\partial s}\dfrac{\partial x}{\partial t} - \dfrac{\partial x}{\partial s}\dfrac{\partial z}{\partial t} & \dfrac{\partial x}{\partial s}\dfrac{\partial y}{\partial t} - \dfrac{\partial x}{\partial t}\dfrac{\partial y}{\partial s} \end{bmatrix} \qquad [4.29]$$

The determinant of the Jacobian becomes:

$$\det\left(\left[J(s,t)\right]\right) = \qquad [4.30]$$
$$\sqrt{\left(\frac{\partial y}{\partial s}\frac{\partial z}{\partial t} - \frac{\partial y}{\partial t}\frac{\partial z}{\partial s}\right)^2 + \left(\frac{\partial z}{\partial s}\frac{\partial x}{\partial t} - \frac{\partial x}{\partial s}\frac{\partial z}{\partial t}\right)^2 + \left(\frac{\partial x}{\partial s}\frac{\partial y}{\partial t} - \frac{\partial x}{\partial t}\frac{\partial y}{\partial s}\right)^2}$$

With $z = 0$, this is the same result as in [4.21].

The integral over the patch of a function $g(s, t)$ depending on the two parameters s and t is performed by transforming this function into another: $g^*(x, y)$, expressed in terms of Cartesian coordinates. Only in special situations is it possible to express $g(s, t)$ in Cartesian coordinates:

$g\ (s,\ t) \rightarrow g^*(x,\ y)$. The integral below often has to be calculated numerically, as its Jacobian cannot be easily calculated analytically:

$$\iint_{\Omega_{Coons}} g^*(x,y)\ dxdy = \int_0^1 \int_0^1 g(s,t) J(s,t)\ dsdt \qquad [4.31]$$

The conductivity matrix for any quadrilateral is calculated numerically. Let k be the thermal conductivity coefficient of a homogeneous and isotropic medium (W m^{-1} K^{-1}) and e its assumed constant thickness (m). The conductivity matrix is written as:

$$K = \iint_{\Omega_{Coons}} ke \begin{bmatrix} \dfrac{\partial \tau}{\partial x} \\ \dfrac{\partial \tau}{\partial y} \end{bmatrix}^T \begin{bmatrix} \dfrac{\partial \tau}{\partial x} \\ \dfrac{\partial \tau}{\partial y} \end{bmatrix} dxdy \qquad [4.32]$$

Using formula [4.30] for the thermal gradient, we obtain the following expression for K, with J as a function of the parameters s and t:

$$\int_0^1 \int_0^1 \begin{matrix} ke \left\{ \begin{bmatrix} \dfrac{\partial x}{\partial s} & \dfrac{\partial y}{\partial s} \\ \dfrac{\partial x}{\partial t} & \dfrac{\partial y}{\partial t} \end{bmatrix}^{-1} \begin{bmatrix} \left[-(1-t) \quad (1-t) \quad t \quad -t\right] \\ \left[-(1-s) \quad -s \quad s \quad (1-s)\right] \end{bmatrix} \right\}^T \\ \left\{ \begin{bmatrix} \dfrac{\partial x}{\partial s} & \dfrac{\partial y}{\partial s} \\ \dfrac{\partial x}{\partial t} & \dfrac{\partial y}{\partial t} \end{bmatrix}^{-1} \begin{bmatrix} \left[-(1-t) \quad (1-t) \quad t \quad -t\right] \\ \left[-(1-s) \quad -s \quad s \quad (1-s)\right] \end{bmatrix} \right\} Jdsdt \end{matrix} \qquad [4.33]$$

The cube element was explicitly developed in Chapter 1, formulas [1.35]–[1.44].

The surface temperature maps of a truncated pyramid with temperatures imposed at nodes 7 and 8: 0 K and 30 K are shown in Figures 4.4–4.6. The average temperature is 16 K. The dissipation calculated according to formula [1.45] is 750 *WK*.

The temperatures at nodes [7 2 3 4 5 6 8 1] are [13 20 9 20 23 13 0 30] K, respectively. The heat flows at the fixed nodes are –50 *W* and 50 *W*. The volume of the element is 0.625 m^3.

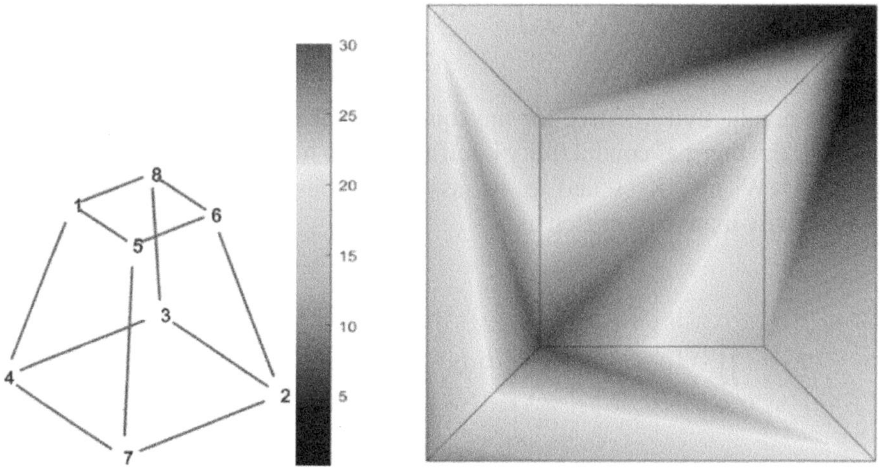

Figure 4.4. *Isotherms on the surface of a truncated pyramid*

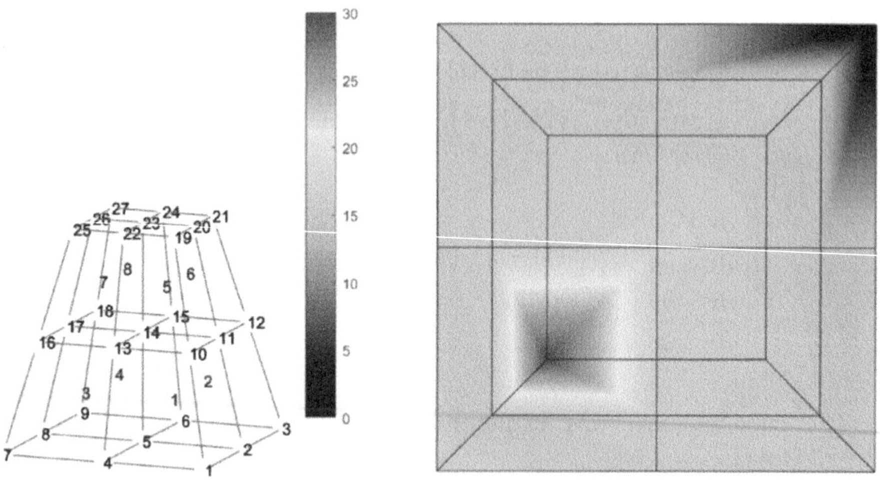

Figure 4.5. *Isotherms with view of the vertices at temperatures of 0 and 30 K*

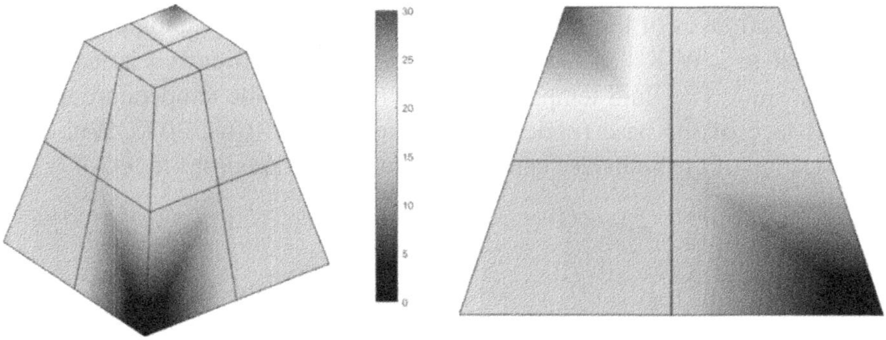

Figure 4.6. *Isotherms on axonometry and section in the plane of nodes 1, 27, 9*

When the domain is subdivided into eight elements, the results shown in Figure 4.6 are obtained. Temperatures of 0 K and 30 K are imposed at nodes 1 and 27. The dissipation, calculated according to formula [1.45], is equal to 353 *WK*.

The heat flows at the fixed nodes are –23.5 *W* and 23.5 *W*. The average temperature is 15.6 K. The volume of the domain is 0.625 m^3. The dissipation ratio for the eight-element model and the single-element model is equal to 353 *WK*/750 *WK* = 0.47. For the cubes in Figures 1.5 and 1.6, the dissipation ratio was 429 *WK*/900 *WK* = 0.48.

4.5. Using fins to cool a solid

We will now consider the case of cooling a solid with three fins. The whole is immersed in a fluid at a temperature lower than that of the solid. The fins are attached to the solid by gluing meshes (see section 2.1.2 and Figure 4.7(a)). The conduction coefficient is equal to 12 W K^{-1} m^{-1}. The domain has three gluing lines connecting edges 9–10, 19–20 and 29–30 to the upper faces of the cubes. In the first example (Figure 4.7), the solid is subject to Dirichlet conditions: temperature set at 320 K on the vertical face connecting nodes 31–34. On the fins, temperatures are imposed at 280 K, 290 K and 300 K on edges 7–8, 17–18 and 27–28. Dissipation is equal to 1,934 *WK*.

To better highlight the function of the cooling fins, the temperatures imposed on the upper edges are replaced by convection conditions on the six large vertical faces of the three fins (Figure 4.8). The convection coefficient h is equal to 2 W K^{-1} m^{-2}, while the conduction coefficient k is equal to 12 W K^{-1} m^{-1}. The fluid temperature is 300 K, while the temperature of the vertical face of the base (nodes 31–34, Figure 4.7(a)) is 320 K. With this modification of the boundary conditions, the dissipation drops to 905 *WK*.

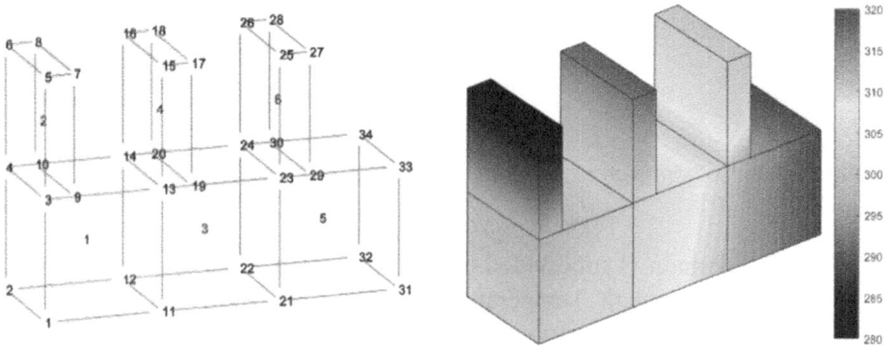

a) b)

Figure 4.7. *Solid with three fins and only imposed temperatures*

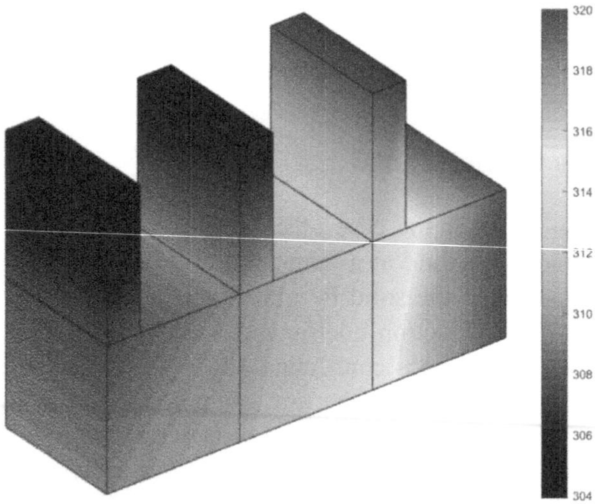

Figure 4.8. *Solid with three glued fins subject to convection*

Part 2

Thermal Radiation

5

Black Bodies

This chapter describes a scene composed of black bodies, whose surfaces radiate according to the fourth power of their temperature, in accordance with the Stefan–Boltzmann law. This is first deduced from Planck's law, which also allows us to verify that the spectrum emitted by the sun and that of a terrestrial scene are almost perfectly separated by a wavelength of 4 μm (Beckers and Beckers 2024). The main quantity of radiation is radiance, a power per unit area and per unit solid angle. Radiation exchanges between the surfaces are determined by the viewing factors. Since the volumes are modeled by isoparametric elements, and the surfaces limiting these elements are Coons patches, we finally show how to calculate the view factor between two Coons patches using optimal ray tracing.

5.1. Planck's law

Planck's law describes the spectral density of electromagnetic radiation emitted by a black body in thermal equilibrium at a given temperature T when there is no net flow of matter or energy between the body and its environment (Planck 1901, 1909).

The *Planck–Einstein* relation links the specific photon energy E to its associated wave frequency ν:

$$E = h\ \nu \qquad [5.1]$$

For a color version of all the figures in this chapter, see www.iste.co.uk/beckers/finite.zip.

The energy of a photon with frequency $\omega = 2\pi\nu$ is given by $E = \hbar\,\omega$, an expression in which we use the *Dirac constant* (or reduced Planck constant) $\hbar = h/(2\pi)$ expressed in the same units as h, namely, in Js.

Planck's constant is:

$$h = 6.62606957 \times 10^{-34}\ \text{Js}$$

The *speed of light* is:

$$c = 299{,}792{,}458\ \text{m s}^{-1}$$

Boltzmann's constant is:

$$k = 1.3806488 \times 10^{-23}\ \text{J K}^{-1}$$

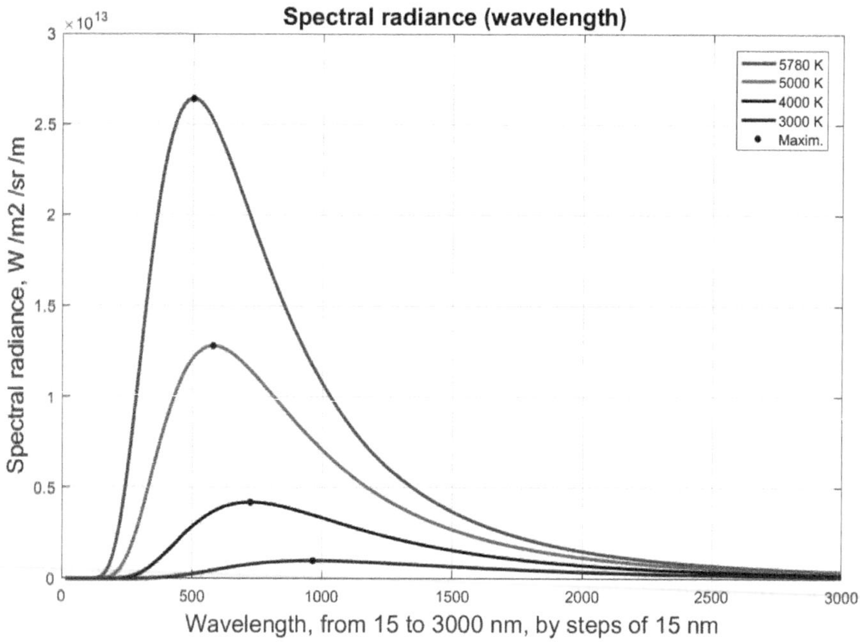

Figure 5.1. *LT (λ) spectral radiances*

Depending on the wavelength, the *spectral radiant power*, or *radiance*, is:

$$L_T(\lambda) = \frac{2hc^2}{\lambda^5} \frac{1}{e^{\frac{hc}{\lambda kT}} - 1} \ \mathrm{Wm^{-2} sr^{-1}\ m^{-1}} \quad [5.2]$$

When *spectral radiance* is integrated over the wavelength range, it becomes *radiance* and is expressed in W m^{-2} sr^{-1}. In the case of a homogeneous diffuse electromagnetic wave emitted on the plane tangent to the surface of the emitter, the result is multiplied by factor π and becomes the blackbody *irradiance* measured in W m^{-2}.

Frequency $\nu = c / \lambda$ is expressed in hertz (Hz or s^{-1}). Spectral radiance $N_T(\nu)$ (Figure 5.1) can also be expressed as a function of frequency (Figure 5.2):

$$N_T(\nu) = \frac{2h\nu^3}{c^2} \frac{1}{e^{\frac{h\nu}{kT}} - 1} \ Wm^{-2} sr^{-1} s \quad [5.3]$$

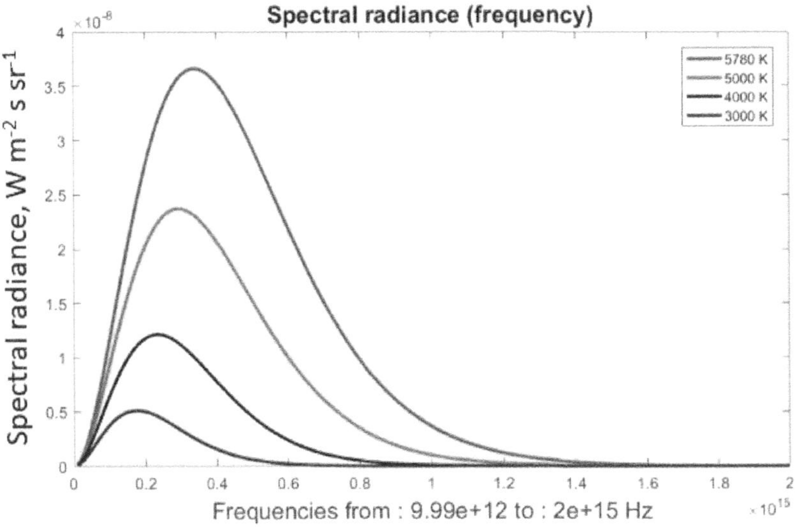

Figure 5.2. *NT(v) spectral radiances*

5.2. Stefan–Boltzmann law

The *Stefan–Boltzmann law* states that the *irradiance* produced by the radiation of a black body is proportional to the fourth power of its temperature:

$$Q = \sigma T^4 \quad \text{W m}^{-2} \text{ with: } \sigma = \ 5.670373\ \times 10^{-8} \text{ W m}^{-2}\text{K}^{-4} \qquad [5.4]$$

The integrals of the spectral radiances introduced in [5.2] and [5.3] give the same irradiance values:

$$\begin{aligned} &\int_0^\infty L_T(\lambda)d\lambda = \int_0^\infty \frac{2hc^2}{\lambda^5}\frac{1}{e^{\frac{hc}{k\lambda T}}-1}d\lambda = \\ &\int_0^\infty N_T(\nu)d\nu = \int_0^\infty \frac{2h\nu^3}{c^2}\frac{1}{e^{\frac{h\nu}{kT}}-1}d\nu \end{aligned} \qquad [5.5]$$

For the first integral, we start with formula [5.2]:

$$\pi\int_0^\infty L_T(\lambda)d\lambda = \pi\int_0^\infty \frac{2hc^2}{\lambda^5}\frac{1}{e^{\frac{hc}{k\lambda T}}-1}d\lambda \qquad [5.6]$$

The wavelength λ is replaced by the dimensionless quantity $x = \dfrac{hc}{\lambda kT}$. This gives:

$$d\lambda = -\lambda^2 \frac{kT}{hc}dx \qquad [5.7]$$

Hence:

$$\int_0^\infty \frac{2hc^2}{\lambda^5}\frac{\pi}{e^{\frac{hc}{k\lambda T}}-1}d\lambda = \int_0^\infty \frac{2hc^2}{\lambda^5}\frac{\pi}{e^x-1}d\lambda = -\pi\int_0^\infty -\frac{2ckT}{\lambda^3}\frac{dx}{e^x-1} \qquad [5.8]$$

We separate the part to be integrated, which depends on x:

$$\pi \int_0^\infty \frac{2ckT}{\lambda^3} \frac{dx}{e^x - 1} = \frac{2\pi (kT)^4}{h^3 c^2} \int_0^\infty \frac{x^3}{e^x - 1} dx \qquad [5.9]$$

The expression [5.9] contains a function of x that can be explicitly integrated, resulting in the expression [5.10]:

$$\int_0^\infty \frac{x^3}{e^x - 1} dx = \frac{\pi^4}{15} \qquad [5.10]$$

Hence:

$$\pi \int_0^\infty L_T(\lambda) d\lambda = \frac{2\pi (kT)^4}{h^3 c^2} \int_0^\infty \frac{x^3}{e^x - 1} dx = \frac{2\pi^5 k^4}{15 c^2 h^3} T^4 \text{ W m}^{-2} \qquad [5.11]$$

We now incorporate the spectral radiance defined in [5.3] by including the π factor:

$$\pi \int_0^\infty N_T(\nu) d\nu = \int_0^\infty \frac{2h\nu^3}{c^2} \frac{\pi}{e^{\frac{h\nu}{kT}} - 1} d\nu = \int_0^\infty \frac{2h}{c^2} \frac{\pi \nu^3}{e^{\frac{h\nu}{kT}} - 1} d\nu \text{ W m}^{-2} \qquad [5.12]$$

As before, the frequency ν is replaced by the dimensionless variable $x = \frac{h}{kT} \nu$:

$$\int_0^\infty \frac{2h}{c^2} \frac{\pi \nu^3}{e^{\frac{h\nu}{kT}} - 1} d\nu = \int_0^\infty \frac{2h}{c^2} \frac{\pi \nu^3}{e^x - 1} \frac{kT}{h} dx = \frac{(kT)^4}{c^2 h^3} \int_0^\infty \frac{2\pi x^3}{e^x - 1} dx \qquad [5.13]$$

Substituting this result into [5.12], we obtain:

$$\pi\int_0^\infty N_T(\nu)\,d\nu = \frac{2(kT)^4}{c^2h^3}\pi\int_0^\infty \frac{x^3}{e^x-1}dx = \frac{2(kT)^4}{c^2h^3}\frac{\pi^5}{15} = \frac{2}{15}\frac{\pi^5k^4}{c^2h^3}T^4 \quad \text{W m}^{-2} \qquad [5.14]$$

According to [5.14] and the relationship defining the Dirac or reduced Planck constant $\hbar = h/(2\pi) = 1.054571783 \times 10^{-34}$ Js, we obtain the *Stefan–Boltzmann constant*:

$$\sigma = \frac{2}{15}\frac{\pi^5k^4}{c^2h^3} \quad \text{or} \quad \sigma = \frac{\pi^2k^4}{60\,c^2\hbar^3} \quad \text{W m}^{-2}\text{K}^{-4} \qquad [5.15]$$

5.3. Short waves and long waves

The Stefan–Boltzmann law establishes the value of the *irradiance* produced by the radiation of a black body:

$$Q = \sigma T^4 \text{ W m}^{-2};\ \sigma = 5.670373\ \times 10^{-8} \text{ W m}^{-2}\text{K}^{-4} \qquad [5.16]$$

The integration of expression [5.16] is performed numerically:

$$Q_{is} = \pi\int_{\lambda_i}^{\lambda_s} L_T(\lambda)\,d\lambda = \pi\int_{\lambda_i}^{\lambda_s}\frac{2hc^2}{\lambda^5}\frac{1}{e^{\frac{hc}{k\lambda T}}-1}d\lambda \text{ W m}^{-2} \qquad [5.17]$$

To calculate the Q_{is}/Q ratio, the integral of relation [5.17] is evaluated at the middle of the n segments used to divide the interval $\lambda_i - \lambda_s$. These values are added together, multiplied by the width of the intervals and by π, and then divided by σT^4.

λ (Nanometer)	Q_{is}/Q	Temperature (K)
0–4,000	0.99027	5,780
4,000–100,000	0.99327	288

Table 5.1. *Evaluation of the Qis/Q ratio*

The wavelength of 4 μm (4,000 nm) almost perfectly separates solar radiation (99% below 4 μm, which are *short waves*) from terrestrial radiation (99% above 4 μm, which are *long waves*).

In the literature, the visible light range is between 380 nm and 760 nm (45% of solar radiation), or between 400 nm and 700 nm (37% of solar radiation). Figure 5.3 shows the spectral radiance of the Sun and a planet such as Earth displayed simultaneously.

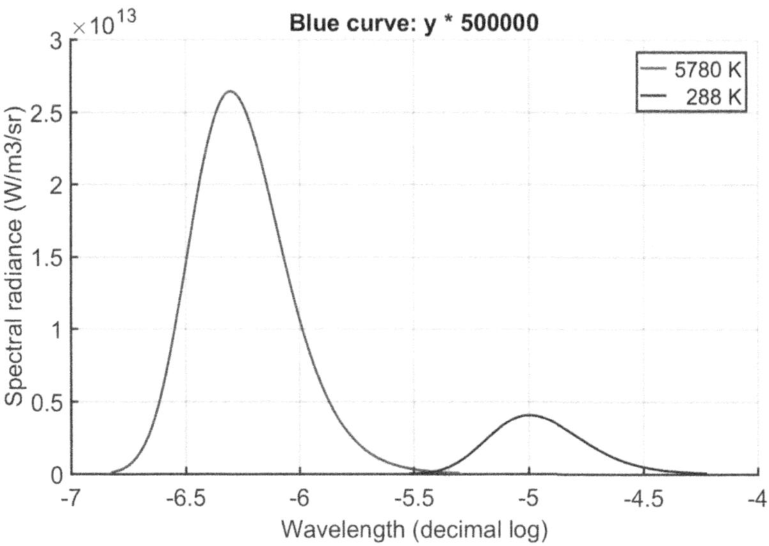

Figure 5.3. *Spectral radiance of the Sun (red) and Earth (blue)*

5.4. Kirchhoff's law

For a patch with a gray, diffuse surface, *Kirchhoff's law* establishes the relationship between emission and reflection:

$$\varepsilon_i = \alpha_i \quad \rho_i = 1 - \varepsilon_i \qquad [5.18]$$

The coefficients introduced in [5.18] represent the following quantities: α_i is the *absorption coefficient*, ε_i is the *emissivity* and ρ_i is the *reflectivity*. All these quantities are dimensionless. In the case of a black body, $\alpha_i = \varepsilon_i = 1$ and $\rho_i = 0$.

5.5. Solid angle

5.5.1. *General procedure*

The area of a sphere with radius r is equal to $4\pi r^2$. The *solid angle* of an object in space relative to a point is given by the area of its projection onto a sphere of unit radius surrounding the point.

The solid angle element $d\omega$ of the area element dA depends on the orientation of the patch and the square of the distance between the viewpoint and the patch:

$$d\omega = \frac{dA \cos\theta_{patch}}{r^2} \qquad [5.19]$$

The variable θ_{patch} represents the angle between the normal to the patch and the line connecting its center of gravity to the observation point. The variable r is the distance between these two points. The concept of solid angle is illustrated in Figure 5.4 by calculating the solid angle ABC of triangle PRQ.

To calculate it, simply follow the procedure for calculating the solid angle of an object in space relative to a point (Beckers and Beckers 2014b).

Procedure:

– project this object onto the unit sphere centered at the chosen point;

– calculate the area of the spherical projection A_p of the object. This is equal to the solid angle sought:

$$A_{S_{x-A}} = A_p = \int_{A_p} d\omega \qquad [5.20]$$

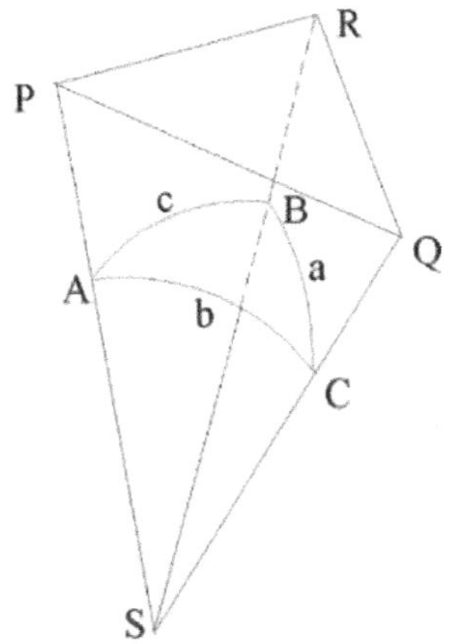

Figure 5.4. *ACB projection of triangle PQR onto the unit sphere centered at S*

NOTE.– The solid angle of the entire space surrounding a point is equal to the area of the unit sphere, that is, 4π. The solid angle above the horizon, that is, corresponding to a hemisphere, is equal to 2π. To visualize the hemisphere in a flat projection, while preserving the areas of the spherical polygons, we use *Lambert's azimuthal projection* (Beckers and Beckers 2014b).

5.5.2. *Area of a spherical polygon*

As shown in Figure 5.4, to calculate the area of a spherical triangle, either the angles at the vertices (A, B, C), or the angles of the sides (a, b, c) are used:

$$
\begin{aligned}
Area_{ABC} = 2E &= A + B + C - \pi \\
tg\frac{E}{2} &= \sqrt{tg\frac{p}{2}\, tg\frac{p-a}{2}\, tg\frac{p-b}{2}\, tg\frac{p-c}{2}} \\
2p &= a + b + c
\end{aligned}
\qquad [5.21]
$$

The area of the spherical polygon is:

$$
Area_{spherical\_polygon} = \left[\sum_{i=1}^{n} \theta_l - (n-2)\pi\right] r^2 \qquad [5.22]
$$

In this expression, the symbols θ_i represent the angles at the vertices (A, B and C for the triangle) of the polygon (expressed in radians), n is the number of sides and r is the radius of the sphere.

5.5.3. *Calculating the solid angle of a triangle*

The projection of triangle PQR onto the unit sphere is the spherical triangle ABC. By definition of this projection, the lengths of the line segments SA, SB and SC are equal to one:

$$
\overrightarrow{SA} = \frac{\overrightarrow{SP}}{\left|\overrightarrow{SP}\right|} \;;\; \overrightarrow{SB} = \frac{\overrightarrow{SQ}}{\left|\overrightarrow{SQ}\right|} \;;\; \overrightarrow{SC} = \frac{\overrightarrow{SR}}{\left|\overrightarrow{SR}\right|} \qquad [5.23]
$$

We denote by A the dihedral angle between faces SAB and SAC (the others being obtained by circular permutation). The sides of this spherical triangle, designated by the lowercase letters a, b and c, are arcs of great circles. Since we are on a unit sphere, the lengths of the arcs a, b and c are equal to the three angles at the vertices S of the spherical pyramid $SABC$ (these angles are measured in radians). The perimeter p of the spherical triangle is denoted by $2p = a + b + c$. The six angles at the vertices A, B, C and those at the sides a, b and c are calculated vectorially from the definition of the pyramid.

The solid angle of triangle *PQR*, viewed from point *S*, is equal to the area of spherical triangle *ABC* constructed on the unit sphere centered at *S*. To calculate its area, we proceed as follows:

– calculate the normals of faces *SPQ*, *SQR* and *SRP* by performing vector products between the edges of the pyramid originating from the vertex:

$$\overrightarrow{n_{SPQ}} = \frac{\overrightarrow{SP} \wedge \overrightarrow{SQ}}{\left|\overrightarrow{SP} \wedge \overrightarrow{SQ}\right|}; \ \overrightarrow{n_{SQR}} = \frac{\overrightarrow{SQ} \wedge \overrightarrow{SR}}{\left|\overrightarrow{SQ} \wedge \overrightarrow{SR}\right|}; \ \overrightarrow{n_{SRP}} = \frac{\overrightarrow{SR} \wedge \overrightarrow{SP}}{\left|\overrightarrow{SR} \wedge \overrightarrow{SP}\right|} \quad [5.24]$$

– calculate the dihedral angles of the faces originating from the vertex by performing scalar products of their normals:

$$\begin{aligned} \cos A &= -\overrightarrow{n_{SPQ}} \cdot \overrightarrow{n_{SRP}} \\ \cos B &= -\overrightarrow{n_{SQR}} \cdot \overrightarrow{n_{SPQ}} \\ \cos C &= -\overrightarrow{n_{SRP}} \cdot \overrightarrow{n_{SQR}} \end{aligned} \quad [5.25]$$

– the solid angle is calculated according to [5.20].

Table 5.2 gives some characteristics of regular polyhedra. Let *Nf* be the number of faces, *n* the number of faces that meet at a vertex of the polyhedron, *m* the number of edges of a face, *Ns* the number of vertices and *As* the solid angle of a face of the solid ({*m*, *n*} is the *Schläfli symbol* of the polyhedron).

	Nf	***n***	**m**	**Ns**	**Solid angle** $As = 4\pi / Nf$	**Angle sum radian, degrees**		**Faces**
Tetrahedron	4	3	3	4	π	$2\pi/3$	120	Triangles
Hexahedron	6	3	4	8	$2\pi/3$	$2\pi/4$	90	Squares
Octahedron	8	4	3	6	$\pi/2$	$2\pi/3$	120	Triangles
Dodecahedron	12	3	5	12	$\pi/3$	$2\pi/3$	120	Triangles
Icosahedron	20	5	3	20	$\pi/5$	$2\pi/5$	72	Pentagons

Table 5.2. *The five regular polyhedra*

5.6. The solar constant

To calculate the *solar constant*, we define the astronomical unit (AU), equal to exactly 149,597,870,700 m, according to the system of astronomical constants in use since 2009. The distance between the Earth and the Sun is 1 AU, or approximately 150 million km. The radius of the Earth, considered a sphere slightly flattened at the poles, is approximately 6,370 km. The radius of the solar sphere is 696,342 km.

The Earth rotates around its axis in just under 24 h (1 day), and around the Sun, in the plane of the ecliptic, with a period of 365.25 days. The axis of the first rotation (which is defined by the two poles) is tilted 23.5° relative to the normal to the ecliptic. This obliquity determines the specific latitudes of the tropics (23.5° relative to the equator) and the polar circles (23.5° relative to the poles and therefore 66.5° relative to the equator).

The surface of the Sun behaves approximately like that of a black body at temperature $T = 5{,}780$ K. According to the Stefan–Boltzmann law [5.4], its irradiance is $\sigma T^4 = 63.29$ MW m^{-2}. The product of this irradiance and the area of the solar sphere (radius: $696{,}340 \times 10^3$ m) is 3.8563×10^{26} MW, which is equal to the product of the solar constant and the area of the sphere with a radius of 150 million km ($4\,\pi\,(150 \times 10^9)^2 = 2.8274 \times 10^{23}$). The resulting solar constant, equal to 1,364 W m^{-2}, is equivalent to the average irradiance captured in the vicinity of the Earth. If this irradiance is distributed, not on a disk perpendicular to the solar radius and with the same radius as the Earth, but on the Earth's sphere, we obtain a value four times smaller, that is, 341 W m^{-2} (Beckers 2012).

5.7. View factor

Remember that, according to the terminology used in CAD, a *patch* is a surface bounded, usually by four sides, and defined in parametric form. This surface can be of any shape; it is not necessarily flat, nor are the sides that bound it necessarily straight lines. The patches in Figure 5.5 are reduced to simple triangles.

The *view factor* (also known as the shape factor) forms the basis of studies on heat transfer by radiation (Sillion and Puech 1994; Beckers 2009; Beckers and Beckers 2012a; Beckers 2023). It is a purely geometric variable, but its definition is based on energy concepts: F_{ij} is the proportion of the total power that leaves element A_i and is received by element A_j.

In formula [5.26], angles θ_i and θ_j correspond to the angles between the vector connecting the two differential elements and the vectors normal to these elements; r is the distance between the two differential elements:

$$F_{ij} = \frac{1}{A_i} \int_{x \in A_i} \int_{y \in Aj} \frac{\cos\theta_i \cos\theta_j}{\pi r^2} V(x,y)\, dydx \qquad [5.26]$$

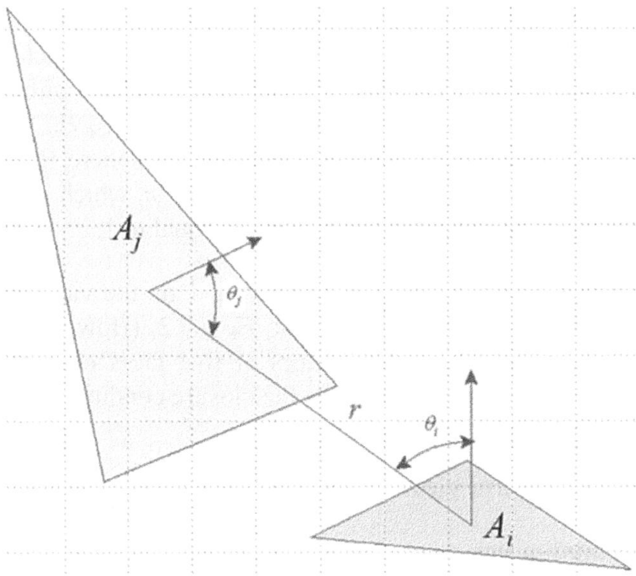

Figure 5.5. *Geometric definition of the view factor*

The symbol V represents the *visibility function*, which is equal to 0 if the ray connecting the two patches is interrupted by an obstacle, otherwise it is equal to 1. Expression [5.26] is symmetric with respect to the two patches (they play the same role). From this, we can deduce the *reciprocity relation*:

$$A_i\, F_{ij} = A_j\, F_{ji} \qquad [5.27]$$

In the following three examples, the view factors are derived directly from the geometric properties of the spaces under consideration.

EXAMPLE 5.1.– Let D be a disk with area A_D and H the hemisphere above it, with area A_H equal to twice that of its base disk: $A_H = 2A_D$.

The view factor F_H of the hemisphere is equal to 1, because all points on the disk see the entire hemisphere. By the reciprocity relation [5.27]: $A_D F_H = A_H F_D$. We deduce that $F_D = A_D F_H/A_H = A_D F_H/(2 A_D) = 0.5$.

EXAMPLE 5.2.– For two concentric spheres, the view factor F_{Si} from the inner sphere to the outer sphere is equal to 1. This is because it cannot see itself, and its elements can see the outer sphere. The view factor F_{Se} of the outer sphere, with area $A\ F_{Se}$, towards the inner sphere (with area A_{Si}) is obtained using the *reciprocity relation* [5.27]: $A_{Si} F_{Se} = A_{Se} F_{Si}$. We therefore have $F_{Si} = A_{Si}/A_{Se}$. The view factor F_{Se} of the outer sphere toward itself is $F_{Se} = 1 - F_{Si}$. This corresponds to the *closure relation,* which states that, in a closed enclosure, the sum of the view factors is equal to 1.

EXAMPLE 5.3.– In a cube with sides measuring 1 m, the view factor of one wall relative to any other is equal to: $F_{ij} = 0.2$ (Howell et al. 2010) (Table 5.3). The sum of the view factors of five faces observed from the sixth is equal to 1 and therefore satisfies the closure condition.

5.8. Calculation of the view factor using the ray tracing method

In the last third of the 20th century, advances in computer technology led to the widespread use of *ray tracing* techniques, particularly in the fields of realistic rendering, lighting and acoustics (Appel 1968). The emergence of this technique accompanied the early developments of CAD (computer-aided design) and computer graphics (Sutherland et al. 1974).

In the standard *Monte–Carlo method*, rays are traced from a point (called the source) and their directions, expressed by two angles in spherical or geographic coordinates (latitude and longitude), are chosen randomly. This is equivalent to tracing randomly within the rectangle of the projection of the sphere onto a square plane (Beckers and Beckers 2014b). This method leads to an excessively high density of rays pointing toward the poles.

The best way to correct this defect is to weight the latitudes according to the distance of the points on the sphere from the equator plane. This is equivalent to performing the projection in the rectangle of the equivalent cylindrical projection (Lambert 1760), also known as the equal-area cylindrical projection. This method, known as *importance sampling*, is a variance reduction technique.

In *stratified sampling* Monte Carlo methods, the ray tracing is no longer taken across the entire domain, but within cells whose union covers the domain (Beckers 2019). The first step is therefore to construct a mesh on the sphere carrying the ray tracing.

To mesh the sphere, it is convenient to define elements delimited by segments of parallels and meridians. The choice of this type of mesh is justified by the fact that spherical coordinates based on polar and azimuthal angles (where the polar angle can be called colatitude, zenith angle, normal angle or inclination angle) or geographic coordinates are widely used to describe the sphere. A direct advantage of this choice is that azimuthal projections centered on the poles are figures of the circle delimited by concentric arcs and segments of radii (Leopardi 2006; Beckers and Beckers 2014a). For these reasons, this method of meshing is preferred over others.

Before addressing the issue of the sphere, let us first examine how to define cells of equal area on a disk. The entire disk is divided into a central disk surrounded by concentric rings, each containing a certain number of cells. For a mesh where all elements have the same area, it is immediately apparent that the sequence of cells differs on the different rings.

To define N equal cells in a unit disk, starting from a central disk composed of a single cell and whose radius is equal to $r_1 = 1/\sqrt{N}$, the calculation can easily be performed in the ring surrounding it. This disk is composed of n cells, so that the disk that is the sum of the inner disk and the outer disk contains ($k_2 = k_1 + n$) cells (where $k_{i+1} = k_i + n$). The radius of this disk is given by $r_{i+1} = r_i \sqrt{k_{i+1}}$. The number of cells added in each ring is arbitrary, provided that the total number of cells does not exceed the value N.

The transfer from the base disk to the hemisphere can be achieved in several ways, but two of them are of particular interest to us: orthogonal projection and equivalent azimuthal projection. For view factors, we use the former (Figure 5.6), while for solid angles, we use the latter (Figure 5.7).

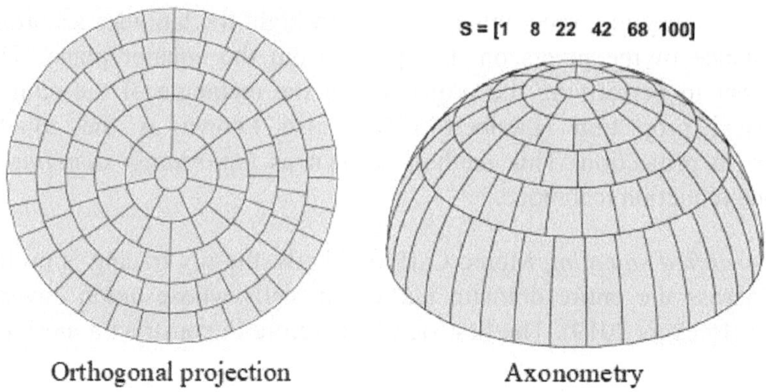

Figure 5.6. *Cells of equal area and equal view factors*

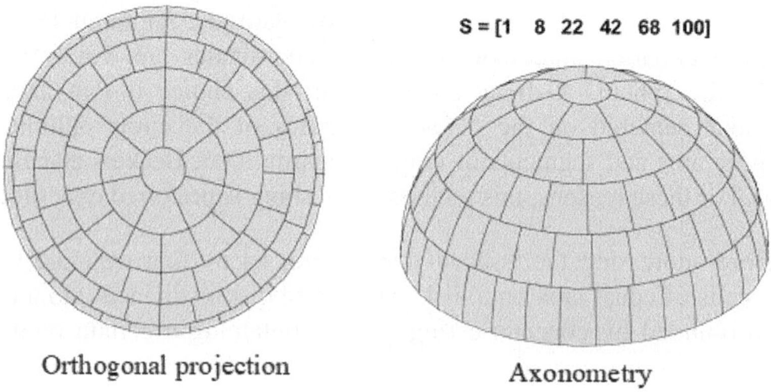

Figure 5.7. *Cells of equal area and equal solid angles*

Once the sequence of cells has been defined on the disk, an inverse azimuthal projection is used to obtain the mesh on the sphere. In the case of an orthogonal azimuthal projection, the relationship between the polar angle θ on the unit hemisphere and the radius in the projection is:

$$r = \sin\theta \qquad [5.28]$$

On the left side of Figure 5.7, we see the orthogonal projection of the hemisphere onto its base. Here, the areas and aspect ratios of the projection are equal. The disadvantage of this choice is the significant distortion of the cells near the base of the hemisphere.

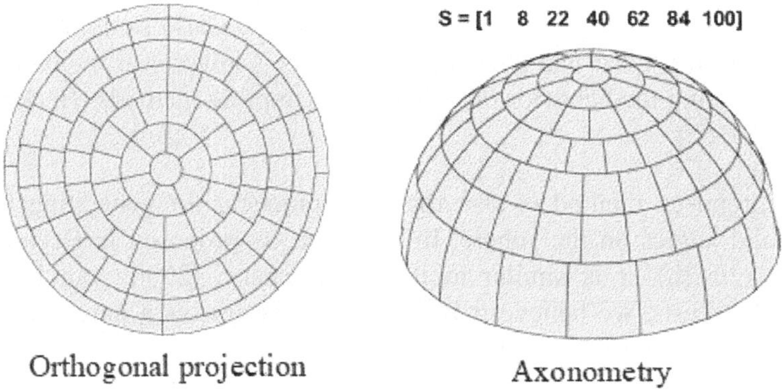

Figure 5.8. *Disk areas, aspect ratios = 1 on the hemisphere*

Since the sequence for filling successive disks is arbitrary, it is possible to impose an additional condition at each stage, for example imposing the aspect ratio of the cells, either in the ring to be inserted on the disk (Figure 5.6) or in its projection onto the hemisphere (Figure 5.8).

In Figure 5.8, the hundred areas of the projection are equal, but with aspect ratios equal to 1 on the hemisphere. Significant distortion of cells near the base of the hemisphere is now avoided.

After creating cells with the same view factors, rays are generated to calculate the view factors of the elements in the scene. The rays are generated in all directions, from an origin point to the scene, and the number of intersections with the elements is counted. The view factor of an element is the ratio between the number of rays *s* that reach it and the total number of rays traced. If the number of rays traced is sufficient, the result tends toward the exact solution (Vujicic et al. 2006).

To define the rays, the first method used is deterministic, for example, the rays pass through the center of each cell. This is the situation shown in the orthogonal projections of the optimized cell sequence [1 15 32 54 80 107 133 151] (Beckers and Beckers 2016b).

In the second ray tracing method, the position in each cell is defined randomly. Since all cells are defined between two latitudes and two longitudes, this procedure is very reliable and easy to implement. This method belongs to the category of stratified Monte Carlo sampling methods (Figure 5.9). It is clear that the density of points is lower near the base of the dome, reflecting the behavior of importance sampling methods (Beckers and Beckers 2016c).

The proposed method is also the most practical for generating rays at equal solid angles on the sphere. In this case, as proposed in Beckers and Beckers (2012b), it is similar to that of Leopardi (2006), but based on comparative tests, we believe it is faster because it uses a purely algebraic procedure. For stratified sampling constructed on cells with the same view factors, the distribution of rays on the hemisphere is uniform (Beckers and Beckers 2016b).

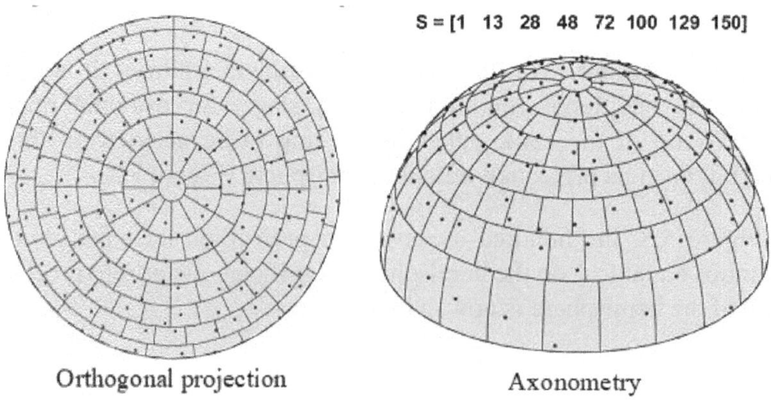

Figure 5.9. *A total of 150 points on the dome*

To calculate the view factors for the faces of the cube, a stratified system should be used by creating a mesh in the face from which the view factors are calculated (Figure 5.10).

With a face of 100 cells and 150 ray tracings onto the hemisphere (Figure 5.9), we obtain the view factors: [0.198 0.198 0.201 0.198 0.0 0.205]. These factors satisfy the closure condition. The minimum is equal to 0.19793, the maximum is equal to 0.20487 and the average is equal to 0.2.

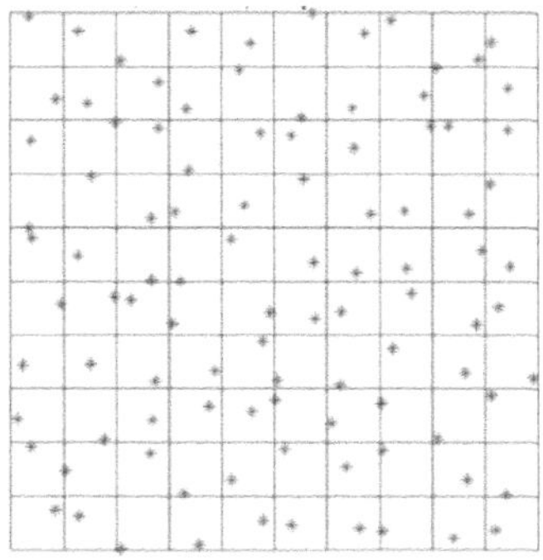

Figure 5.10. *A total of 100 points selected at random using stratified sampling*

The ray tracing method has the enormous advantage of allowing surface-to-surface view factors to be calculated correctly. To do this, it is sufficient to create a mesh on the patch from which the rays are traced.

At each of the 100 points on the base of the cube shown in red (Figure 5.10), we trace 150 oriented rays (as shown in Figure 5.9). When we run this program several times with the same data, the points of the dome cells change position within their cell; the points of the base patch (Figure 5.10) also change position within their cell. The result, therefore, varies slightly from one calculation to another. The view factors of the faces from which the rays originate are zero. They average 0.2 for the other five faces. These factors verify the closure condition (Table 5.3).

View factors of the cube faces
0.0, 0.2, 0.2, 0.2, 0.2, 0.2,
0.2, 0.0, 0.2, 0.2, 0.2, 0.2,
0.2, 0.2, 0.0, 0.2, 0.2, 0.2,
0.2, 0.2, 0.2, 0.0, 0.2, 0.2,
0.2, 0.2, 0.2, 0.2, 0.0, 0.2,
0.2, 0.2, 0.2, 0.2, 0.2, 0.0

Table 5.3. *Exact view factors of the six faces of the cube*

The random creation of points in a 1 × 1 square can be used to generate a Coons patch in the parametric coordinate system corresponding to a surface element of any polyhedron (Figure 5.11).

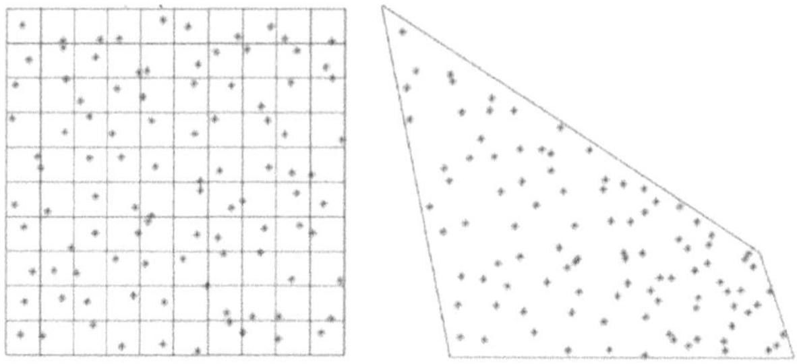

Figure 5.11. *A total of 100 points on a square or a quadrilateral*

6

The Sun and Short Waves

This chapter describes surfaces that appear colored, matte, specular or translucent to the human eye. The almost perfect separation between the shortwave spectrum of the Sun and the longwave spectrum emitted by surfaces whose temperature is close to the average temperature on Earth makes shortwave radiation a problem that is practically decoupled from the diffusion equation studied in the first part of this book. In general, the Sun is the only source. Under the Earth's atmosphere, the sky is considered a secondary source. These two sources are external to the scene under study. The scene modifies the light field through multiple reflections, but does not provide any additional sources, since the Earth's emission is entirely produced in long waves. Therefore, solar radiation acts solely as an external load, which can be calculated independently of the finite elements.

If the observed scene is located on the Earth's surface at a given latitude, we must first express the apparent path of the Sun in the sky for the period under study. After a brief review of ellipses, we introduce Keplerian orbits and time counting, which allow us to determine the position of the Sun based on latitude and time. The attenuation of solar radiation by the atmosphere is then taken into account, and the radiosity equation is introduced in order to calculate the infinite number of diffuse reflections that occur within the scene. Finally, the concept of extended view factor is recalled. It allows other types of reflection to be taken into account in the same formalism, starting with specular reflection, which occurs on mirrors.

For a color version of all the figures in this chapter, see www.iste.co.uk/beckers/finite.zip.

6.1. The ellipse and its anomalies

The concepts discussed below are standard; they are covered in textbooks on synthetic geometry, analytical geometry and orbital mechanics (Prussing and Conway 1993).

The geometry of the ellipse can be described in various ways.

1) The ellipse is defined by its two semi-axes a and b (Figure 6.1). It is obtained from the circle by an affine transformation consisting of a change of scale along the direction of one of its two axes.

In Cartesian coordinates, the equation of a circle with radius a is written as:

$$x^2 + y^2 = a^2 \qquad [6.1]$$

By dividing the variable y by the factor b/a, we obtain the equation of the ellipse in Cartesian coordinates:

$$\frac{x^2}{a^2} + \frac{y^2}{a^2} / \left(\frac{b}{a}\right)^2 = 1; \quad \frac{x^2}{a^2} + \frac{y^2}{b^2} = 1 \qquad [6.2]$$

2) The ellipse can also be seen as a trajectory for which the distance to a given point varies between the values α and β, called the *apocenter* and *pericenter*, which connect the ends of its major axis to one of its foci (Figure 6.1).

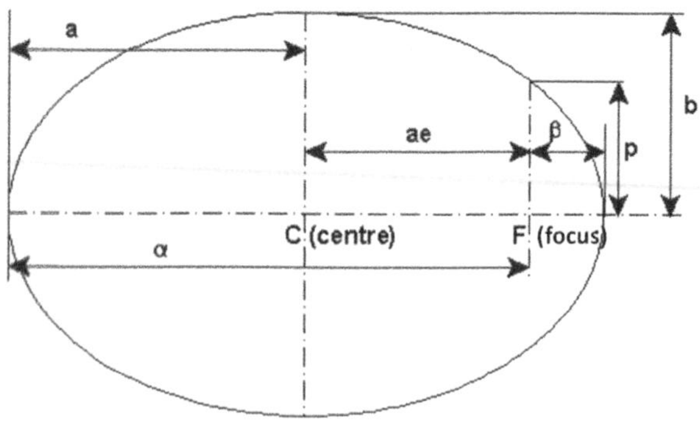

Figure 6.1. *Ellipse parameters*

These two pairs of data are linked by the relationships:

$$a = \frac{\alpha + \beta}{2}; \; b = \sqrt{\alpha\beta} \tag{6.3}$$

The inverse relationships are given as:

$$\alpha = a + \sqrt{a^2 - b^2}; \; \beta = a - \sqrt{a^2 - b^2} \tag{6.4}$$

3) The ellipse can also be defined using two new variables used in conic section theory: e and p. The *eccentricity* e of a conic section defines the relative position of the focus on the semi-major axis. The distance between the focus and the center is equal to $e\ a$. For the ellipse, the eccentricity is naturally between zero and one: $0 \le e < 1$.

Let us write the equation of the ellipse in polar coordinates (ρ, v) [6.5]. The variable ρ represents the distance from a point on the ellipse to the focus. The angle v is called the *true anomaly*. The term p corresponds to the radius ρ when the anomaly v is equal to 90°; this is the parameter of the conic section, also called the *semilatus rectum* (the *latus rectum* is the chord of the conic section perpendicular to its major axis and passing through the focus):

$$\frac{1}{\rho} = \frac{1 + e\cos v}{p} \tag{6.5}$$

This gives us the relationships linking p and e to α and β:

$$e = \frac{\alpha - \beta}{\alpha + \beta}; p = \alpha\ (1 - e) = \beta\ (1 + e); p = \frac{2\alpha\beta}{\alpha + \beta} \tag{6.6}$$

Using relations [6.3]–[6.6], we find another useful relation, this time linking the semi-axes to the eccentricity:

$$\frac{b}{a} = \sqrt{1 - e^2} \; or \; e = \sqrt{1 - \frac{b^2}{a^2}} \tag{6.7}$$

Knowing that the circle with radius a is circumscribed around the ellipse with axes $2a$ and $2b$, we can easily calculate the area of the ellipse. It is equal to that of the circle, multiplied by the ratio b/a, that is, πab.

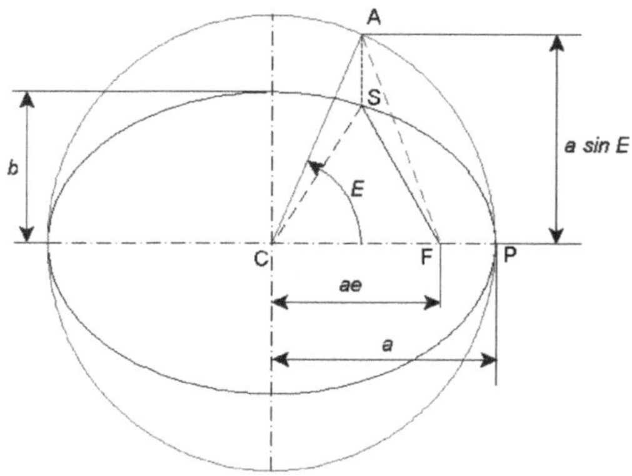

Figure 6.2. *Eccentric anomaly E*

Consider a sector of the circle, for example, sector *CPA* (Figure 6.2), with angular aperture E. Its area is equal to $Ea^2/2$. If we subtract triangle *CFA*, the area of the truncated sector is equal to:

$$Area\ \text{FPA} = \frac{Ea^2}{2} - \frac{1}{2} ae\, a \sin E \qquad [6.8]$$

Scaled along the vertical axis, this element becomes an ellipse sector *FPS* with an area equal to:

$$Area\ \text{FPS} = \frac{ab}{2}(E - e\sin E) \qquad [6.9]$$

Angle E is called the eccentric anomaly.

To divide the area of the ellipse into equal quarters starting from one of its foci, we must calculate eccentric anomalies E_i such that:

$$\frac{\pi ab}{n} = \frac{ab}{2}\left(\left(E_{i+1} - e\sin E_{i+1}\right) - \left(E_i - e\sin E_i\right)\right) \qquad [6.10]$$
$$i = 0, n \text{ and } E_n = E_0 + 2\pi$$

Starting the cut at the end of the major axis on the side of the pericenter (point P, Figure 6.2), we obtain the cut points:

$$E_1 - e\sin E_1 = \frac{2\pi}{n};\ E_2 - e\sin E_2 = \frac{4\pi}{n};\ ... \qquad [6.11]$$

The problem seems easy at first glance, but we need to solve the transcendental equation [6.12], for which there is no explicit solution:

$$M = \left(E - e\sin E\right) \qquad [6.12]$$

In this relationship, the left-hand member M represents an angle: the *mean anomaly* is the angle that would correspond to the division of the circle into identical quarters (twelve quarters, $n = 12$ in Figure 6.3):

$$0;\ \frac{2\pi}{n};\ \frac{4\pi}{n};\ ...\ ;\ 2\pi \qquad [6.13]$$

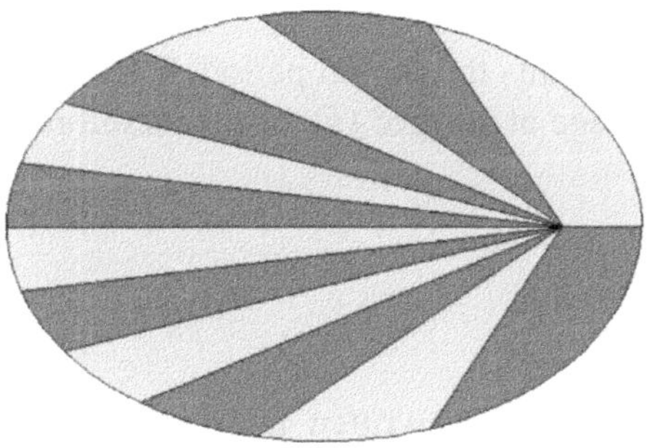

Figure 6.3. *Twelve equal-area quadrants of an ellipse*

The following is a summary of the definitions of three anomalies.

The *true anomaly* is the angle formed by the radius vector drawn from the center of attraction to the orbiting object with the line of apsides (perigee and apogee for the Earth's orbit).

The *mean anomaly* is the angle formed by the radius vector drawn from the center of attraction to an object revolving in a uniform motion and the line of apsides.

The *eccentric anomaly* is the angle measured between the major axis and the radius from the center of the ellipse to the point on the circumscribed circle that has the same abscissa as the current point.

6.2. Keplerian orbits

If we adopt the so-called two-body simplification, introduced by Kepler based on his observations of planetary motion (Prussing and Conway 1993), the trajectory of a planet around the Sun or of a satellite around a planet is described by an ellipse. According to Kepler's second law, the satellite sweeps equal areas in equal times. This means that the time spent on a segment of the trajectory is proportional to the period of the orbit, as is the area swept on the surface of the ellipse. For each segment of the ellipse in the quarters of Figure 6.3, the time is therefore the same (Figure 6.4):

$$\frac{\Delta t}{T} = \frac{sector\ area\ \mathrm{FPS}}{ellipse\ area} \qquad [6.14]$$

The variable T represents the orbital period. Starting from [6.12], the time elapsed since the passage at perigee ($E = 0$) is therefore calculated using the following formula:

$$t = \frac{T}{2\pi}\left(E - e\sin E\right) \qquad [6.15]$$

The relationship [6.15] is rewritten in [6.16] in the form of an angle, without explicitly involving the period:

$$M = \frac{2\pi t}{T} = \left(E - e\sin E\right) \qquad [6.16]$$

M is the angle traveled during the same period of time on a circular orbit with the same period. Since the motion is uniform on circular orbits, this definition provides a kind of clock whose dial is the period of the orbit.

In section 6.1, angle M was referred to as the *mean anomaly*. Equation [6.12], reproduced in [6.16], is the famous *Kepler's equation*. Knowing M and e, we seek to calculate E. Since there is no explicit solution to this equation, it must be solved numerically, using an iterative method.

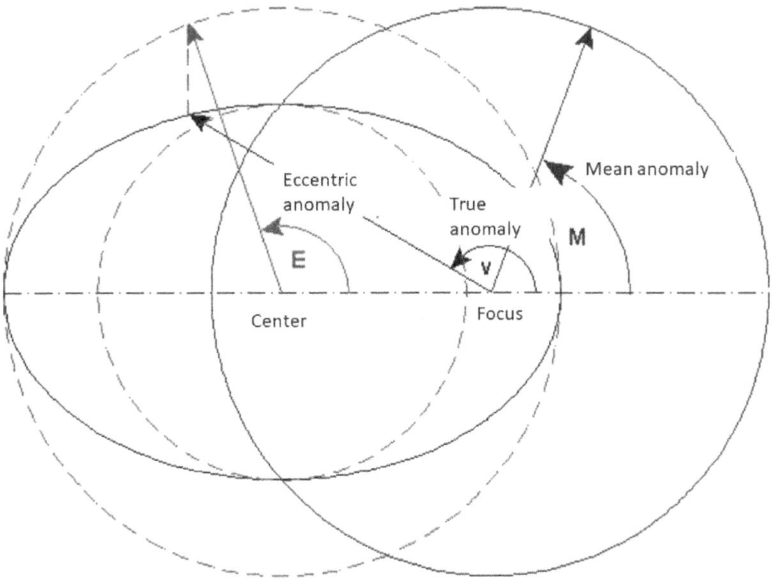

Figure 6.4. *The three anomalies*

If the orbit is slightly elliptical, it is useful to calculate the offset between the *mean anomaly,* which gives the time scale, and the *true anomaly,* which gives the position. This is the case for the Earth's orbit.

The Earth's movement around the Sun is described by an orbit with eccentricity:

$$e_{Earth} = 0.016710219 \qquad [6.17]$$

We can calculate the offset between a point moving along the Earth's actual orbit and a point moving along a circular orbit with the same period. The blue circle in Figure 6.4, corresponding to uniform motion (variable M), is graduated in hours and minutes. With the eccentricity of formula [6.17], the offset reaches a maximum of eight minutes (Figure 6.5). In this graph, the angle of offset is calculated in the time system, meaning that the 360° of the complete circle is replaced by 24 h = 24 × 60 min. Four minutes on this graph therefore correspond to 1°. The only data in this problem is the eccentricity of the orbit.

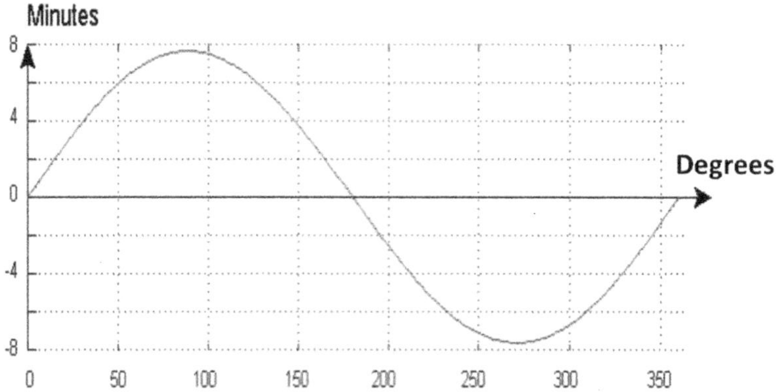

Figure 6.5. *Deviation in minutes from the circular orbit*

According to [6.14], the mean anomaly M corresponds to the motion on a circular orbit with the same period as the elliptical orbit:

$$M = \left(E - e\sin E\right) \qquad [6.18]$$

It was Kepler who first raised this problem, now known as *Kepler's problem*. The problem involves finding the polar coordinates of a planet at a given time, or, equivalently, given the mean anomaly of a planet, deducing its true anomaly. Kepler provided an approximate solution. Other mathematicians such as Wallis, Newton, La Hire, Cassini and Lalande provided more complete solutions.

The three anomalies are shown in Figure 6.4, where the elliptical orbit is shown in black. They are linked by purely geometric relationships. We have just seen the relationship between the *mean anomaly* and the *eccentric anomaly*. Here is the relationship between the latter and the *true anomaly*:

$$tg\frac{v}{2} = \sqrt{\frac{1+e}{1-e}}\, tg\frac{E}{2} \qquad [6.19]$$

We can also calculate the relationship between these two anomalies by considering the two components of the position. Let us return to Figure 6.2

and assume that the abscissa is directed along the major axis and that the origin of the axes is at the focus:

$$\begin{aligned} x &= \rho\cos v = a(\cos E - e) \\ y &= \rho\sin v = b\sin E = a\sqrt{1-e^2}\sin E \end{aligned} \quad [6.20]$$

This leads to a new expression for the distance ρ to the focus, equivalent to [6.5], but this time as a function of the eccentric anomaly:

$$\rho = \alpha(1 - e\cos E) \quad [6.21]$$

This last expression allows us to rewrite formula [6.19] linking the two anomalies in another form; the distances ρ and a have been eliminated from the expression:

$$\begin{aligned} \cos v &= \frac{\cos E - e}{1 - e\cos E} \\ \sin v &= \frac{\sqrt{1-e^2}\sin E}{1 - e\cos E} \end{aligned} \quad [6.22]$$

Equation [6.22] is equivalent to [6.19], but has the advantage of removing any ambiguity regarding angles. These two sections are based on the study of Beckers and Beckers (2010a, 2010b).

6.3. The position of the Sun

The position of the Sun varies over time according to two main periodicities. The first, due to the Earth's rotation on its axis, is 24 h, and the second, due to the Earth's rotation around the Sun, is 1 year. However, while the first movement can be considered uniform, the same cannot be said for the second. The two main causes of its non-uniformity are the tilt of the Earth's axis of rotation relative to its orbital plane and the elliptical shape of its orbit.

To locate a planet in its orbit, you first need to know where it was at a specific time and then use its equation of motion and all the parameters that define its orbit.

The usual date references are either the *Julian Day* (*JD*) or the *Modified Julian Day* (*MJD*). There are websites where you can find their values. Here, we will use a definition close to *J2000* (Schlyter 1994).

The Julian day corresponding to January 1, 2000, at noon is:

$$JD = 2{,}451{,}545.000000\ (UT)$$

If that day is the first day of the numbering, it is linked to the Julian day, or modified Julian day, by the relationship:

$$d = JD - 2451543.5 = MJD - 51543.0 = J2000 + 0.5 \qquad [6.23]$$

The date reference is the day number starting from January 1, 2000, at 00:00 Universal Time (UT). There is a difference of 0.5 days, because d starts at 00:00 while J2000 starts at noon. The day number d is calculated using the following formula:

$$d = 367Y - \frac{7}{4}\left[Y + \frac{Mo + 9}{12}\right] + \frac{275Mo}{9} + D - 730{,}530 \qquad [6.24]$$

This equation includes the following variables: Y is the four-digit year, Mo is the month and D is the day of the month. All divisions are performed by rounding the results down to the nearest whole number. For January 1, 2000, the result is equal to 1.

After calculating the day number, we use the ephemerides to calculate the position of the object (planet or satellite being studied).

Sun Ephemerides		
Longitude of perihelion	w	$282.9404° + 4.70935° \times 10^{-5}$ days
Eccentricity	e	$0.016709 - 1.151 \times 10^{-9}$ days
Mean anomaly	M	$356.0470° + 0.9856002585°$ days
Mean longitude of the Sun	$L_{average}$	$w + M$
Obliquity of the ecliptic	ε	$23.4393° - 3.563° \times 10^{-7}$ days

Table 6.1. *Solar ephemerides (Schlyter 1994)*

Based on the day number as defined in [6.24], the ephemerides of the Sun are given in Table 6.1. The longitude of the perihelion is measured from the vernal point, or vernal equinox point, which corresponds to the *spring equinox*. This point is located on the line of intersection of the plane of the equator and the plane of the ecliptic.

For $d = 3{,}728.5$ (March 16, 2010 at 12.00 *UT*), this gives:

– $w = 283.1160°$;

– $e = 0.01670471$;

– $M = 70.8576°$;

– $L_{average} = 353.973°$;

– $\varepsilon = 23.4380°$.

The *mean anomaly* M corresponds to the motion on a circular orbit with the same period as the elliptical orbit:

$$M = E - e\sin E \qquad [6.25]$$

Using the value of M (Table 6.1) and Kepler's equation [6.25], we calculate the *eccentric anomaly* E, expressed in degrees:

$$E = 71.7666° = 1.2526\ radians \qquad [6.26]$$

Equation [6.20], where a is taken to be equal to 1 (1 AU), provides the Cartesian coordinates x and y of the Sun's position in the coordinate system defined by the axes of the elliptical orbit:

$$\begin{aligned} x &= \rho\cos v = 0.2962 \\ y &= \rho\sin v = 0.9497 \end{aligned} \qquad [6.27]$$

This position can be converted into distance ρ and true anomaly v. We find:

$$\begin{aligned} \rho &= 0.9954\ UA \\ v &= 72.0817° \end{aligned} \qquad [6.28]$$

Adding the longitude of the perihelion $w = 283.12°$ (Table 6.1), we obtain the longitude of the Sun $L_{average} = (w + v) = 355.2°$. From this, we can deduce the position of the Sun in Cartesian coordinates:

$$\begin{aligned} x &= \rho \cos\ (w + v) \\ y &= \rho \sin\ (w + v) \\ z &= 0 \end{aligned} \qquad [6.29]$$

Until now, we have been working in the plane of the Earth's orbit, or the plane of the ecliptic. To find your bearings on Earth, you need to use a system based on the Earth's equatorial plane and the axis of the poles. This requires a rotation of 23.5° (more precisely, an angle of $\varepsilon = 23.44°$ (Table 6.1):

$$\begin{bmatrix} x_{equat} \\ y_{equat} \\ z_{equat} \end{bmatrix} = \begin{bmatrix} 1 & 0 & 0 \\ 0 & \cos\varepsilon & -\sin\varepsilon \\ 0 & \sin\varepsilon & \cos\varepsilon \end{bmatrix} \begin{bmatrix} x \\ y \\ z \end{bmatrix} \qquad [6.30]$$

We can then calculate the *right ascension* φ_{AD} and *declination* δ of the Sun. These two spherical coordinates are similar to the longitude and latitude used to locate points on Earth:

$$\begin{aligned} \varphi_{AD} &= \text{atan2}\ (y_{equat}\ ,\ x_{equat}) \\ \delta &= \text{atan2}\ (z_{equat}\ ,\ \sqrt{x_{equat}^2 + y_{equat}^2}\,) \end{aligned} \qquad [6.31]$$

In the previous formula and in [6.37], the selected Matlab® functions ensure that the angles are calculated correctly.

6.4. Sidereal time and hour angle

On the celestial sphere, the planes of the equator and the ecliptic intersect along a straight line whose ends on the terrestrial sphere are called nodes. During its apparent motion, the Sun crosses these two points, one when passing from the northern hemisphere to the southern hemisphere, which is the descending node; the other when passing from the southern hemisphere to the northern hemisphere, which is the ascending node. The latter is the *vernal point*.

By definition, sidereal time is zero when the vernal equinox passes through the meridian plane of the location in question, and it increases by one sidereal hour each time the Earth rotates 15° relative to the vernal point.

In a way, *sidereal time* measures the movement of the celestial sphere in a given location relative to the local meridian.

Despite its name, sidereal time is not time in the usual sense, but rather the measurement of the angle between the vernal point and the meridian plane.

Greenwich Mean Solar Time (*GMSTO*) is calculated from the mean longitude of the sun $L_{average} = w + M$ (Table 6.1), which gives the solar time:

$$GMSTO = \frac{w+M}{15} + 12 = \frac{L_{average}}{15} + 12 \qquad [6.32]$$

In our example, using the ephemerides table (Table 6.1) and [6.31], we calculate Greenwich Mean Time: $GMSTO = 11.5654$. We subtract 360° or 24 h if necessary to keep the angle between 0° and 360° or the time between 0 and 24 h.

Local sidereal time (*SIDTIME)* is calculated by adding to Greenwich Mean Time (*GMSTO*) the time in Coordinated Universal Time (*UT*, which is the local time minus the time zone offset) and the longitude φ of the point being studied, counted positively toward the east and expressed in hours:

$$SIDTIME = GMSTO + UT + longitude \qquad [6.33]$$

The hour angle φ_{HA} indicates in hours the direction in which the Sun is located. It is equal to the sidereal time *SIDTIME* minus the right ascension φ_{AD}. All angular measurements are expressed in hours:

$$\varphi_{HA} = SIDTIME - \varphi_{AD} \qquad [6.34]$$

The position of the Sun, as a function of the hour angle φ_{HA} and the declination of the Sun δ, is therefore:

$$\begin{aligned} x_{sun} &= \cos\varphi_{HA}\cos\delta \\ y_{sun} &= \sin\varphi_{HA}\cos\delta \\ z_{sun} &= \sin\delta \end{aligned} \qquad [6.35]$$

All that remains is to express this position in local coordinates defined by the local horizontal plane tangent to the Earth's surface at latitude l_a. This includes the south and east directions. It is perpendicular to the vertical axis pointing toward the local zenith:

$$\begin{bmatrix} x_L \\ y_L \\ z_L \end{bmatrix} = \begin{bmatrix} \sin l_a & 0 & -\cos l_a \\ 0 & 1 & 0 \\ \cos l_a & 0 & \sin l_a \end{bmatrix} \begin{bmatrix} x_{sun} \\ y_{sun} \\ z_{sun} \end{bmatrix} \qquad [6.36]$$

We can now calculate the *azimuth* (angle measured from north) and the *height* of the sun:

$$\begin{aligned} azimuth &= \text{atan2}\,(y_L, x_L) + \pi \\ height &= \arcsin\,(z_L) \end{aligned} \qquad [6.37]$$

This concludes the calculation of the Sun's position at a point with *longitude* and *latitude* coordinates l_a at the specified time (variable *UT*).

Performing the calculations for Brussels (4.3° *East*, 50.8° *North*), on March 16, 2010 at 1 p.m., the simulation results are, for right ascension: $\varphi_{AD} = -0.2573325$ h or $\varphi_{AD} = 23$ h 44.56', and for the declination: $\delta = -1°40.3'$ [6.31]. We also obtain [6.37]: *azimuth* = 2.63°, *height* = 37.3°. The same values are obtained on websites dedicated to these calculations (Beckers and Beckers 2010a): $\varphi_{AD} = 23$ h 44.56', $\delta = -1°40.3'$, *azimuth* = 2.7° and *height* = 37.5°.

6.5. Position of the Sun based on latitude and time

The declination of the Sun (angle between the direction of the Sun and the plane of the equator) is calculated using the formula from Campbell and

Norman (1998). This uses the inclination of the Earth's axis relative to the plane of the ecliptic ε = 23.45° or 0.4093 radians. Similar approaches can be found in many references.

In the first two formulas, the arguments of the trigonometric functions are expressed in radians. The variable J denotes the number of the day of the year in the calendar (1 for January 1):

$$\sin\delta = \sin\varepsilon \sin\frac{\pi}{180}\begin{pmatrix} 278.97 + 360\dfrac{J}{365.25} + \\ 1.917\sin\left(\dfrac{\pi}{180}\left(356.6 + 360\dfrac{J}{365.25}\right)\right) \end{pmatrix} \quad [6.38]$$

$$\sin\delta = \sin\varepsilon \sin\begin{pmatrix} 4.8689 + 0.0172J + \\ 0.0334\sin(6.2238 + 0.0172J) \end{pmatrix} \quad [6.39]$$

In [6.38], the arguments are expressed in degrees as in Campbell and Norman (1998). This formula takes into account the ellipticity of the Earth's orbit:

$$\sin\delta = \sin\varepsilon \sin\begin{pmatrix} 278.97 + 360\dfrac{J}{365.25} + \\ 1.917\sin\left(356.6 + \dfrac{360}{365.25}J\right) \end{pmatrix} \quad [6.40]$$

The zenith angle ψ of the sunbeam is given by:

$$\cos\psi = \sin l_a \sin\delta + \cos l_a \cos\delta \cos\left[15(t - t_0)\right] \quad [6.41]$$

In this expression, the variable l_a denotes latitude, t denotes time (defined in hours) and t_0 denotes the time at solar noon (when the Sun passes its highest point).

Every day, the minimum value of the zenith angle (maximum height of the Sun) is obtained at noon:

$$\psi = l_a - \delta \quad [6.42]$$

The absolute minimum occurs at the summer solstice in the hemisphere for latitudes outside the intertropical zone. The half-length of the day measured in hours is given by:

$$h_s = \frac{12}{\pi}\arccos\left(\frac{-\sin l_a \sin\delta}{\cos l_a \cos\delta}\right) = \frac{12}{\pi}\arccos\left(-\tan l_a \tan\delta\right) \qquad [6.43]$$

Since the declination of the Sun lies between the latitudes of the two tropics: $-\varepsilon \leq \delta \leq \varepsilon$, the use of this relationship requires some precautions for points located at latitudes outside those of the polar circles ($l_a = \pm$ (90–ε) $= \pm$ 66.55°). Note that at these latitudes, we are outside the domain of validity of the model of Liu and Jordan (1960).

The azimuth *AZ* of the Sun is equal to:

$$\cos AZ = \frac{-\left(\sin\delta - \cos\psi \sin l_a\right)}{\cos l_a \sin\psi} \qquad [6.44]$$

These formulas will enable the calculation of irradiance at any given moment and at any point on the globe.

6.6. Attenuation of solar radiation by the atmosphere

The aim is to show how to calculate the direct and diffuse components of solar irradiance on a clear day and to deduce the maximum energy received at any point on the globe within a given time interval (Beckers and Beckers 2011). This approach ignores two phenomena: cloud cover, which alters the ratio between the direct and diffuse components of sunlight, and reflection, which adds a third component often referred to as albedo. The values examined here are only meaningful in terms of instantaneous power or energy integrated over a perfectly sunny day in the absence of any reflective surfaces. However, these values serve as a reference for more advanced calculations. They show how the solar irradiance arriving above the atmosphere (the solar constant) is modified by the atmosphere and by the obliquity of the solar flux falling on the surface under study. This surface can be a measuring instrument (pyranometer or pyrheliometer), a solar panel, a facade, or even a window, located anywhere on the globe between the equator and the poles.

6.6.1. *Atmospheric pressure*

The calculation of atmospheric attenuation of solar radiation is based on the assessment of *atmospheric pressure* and the *optical air mass*. Atmospheric pressure is calculated based on altitude h expressed in meters:

$$\frac{p_a}{p_0} = e^{-\frac{h}{8{,}200}} \qquad [6.45]$$

At sea level, the pressure is: $p_0 = 101{,}325$ Pascals (or 1 013.25 hPa, or 1,013.25 millibars, or 1.013 bar). The standard temperature there is 15°C, or 288 K. In the troposphere, which extends from approximately 0 to 11 km, the temperature decreases linearly by about 6.5 K per km. To calculate atmospheric pressure (Piedallu and Gégout 2007) we use the formula proposed by the International Civil Aviation Organization:

$$\frac{p_a}{p_0} = e^{-\frac{h}{8{,}200}} \frac{p_a}{p_0} = \left(1 - \frac{6.5}{288}\frac{h}{1{,}000}\right)^{5.256} \qquad [6.46]$$

The results obtained in [6.44] and [6.45] differ slightly and are even almost identical for altitudes below 4,000 m (for h = 3,000 m, [6.45] gives: 0.6936 and [6.46]: 0.6918).

6.6.2. *The optical air mass*

The *optical air mass* is a measurement taken at sea level of the length of the path traveled through the atmosphere by light rays from a celestial body; it is expressed as a multiple of the length of the path corresponding to a light source located at the zenith.

According to the meteorology glossary published by the American Meteorological Society (AMS), for zenith distances (remember that the zenith distance is the angle between the ray and the local vertical) up to approximately 70°, the optical air mass m_o is approximately equal to the secant of the angle defining the zenith distance of the given celestial body [6.44]. For a more accurate calculation, the refraction of the light ray must be taken into account. To obtain a representative value at high altitude, these values must be multiplied by the ratio between the actual atmospheric pressure and the pressure at sea level.

The second formula (Campbell and Norman 1998) gives the optical air mass as a function of the zenith angle ψ of the solar ray and atmospheric pressure. This formula corresponds to an atmosphere modeled as a layer of constant thickness placed on a plane tangent to the Earth's surface. The optical air mass is related to altitude by one of the formulas [6.45] or [6.46] for calculating atmospheric pressure:

$$m_0 = \sec\psi;\ m = \frac{p_a}{p_0} m_0 \qquad [6.47]$$

To calculate optical air mass, it is better to use the simplest formula [6.43], known as the secant formula. Other formulas do not offer any significant improvement and have the disadvantage of requiring additional physical data – the height of the atmosphere – which is difficult to determine.

We can calculate m directly as a function of altitude using combinations of the formulas defined for the two entities: atmospheric pressure and optical mass number. By combining formulas [6.45] and [6.47], we obtain the expression:

$$m = e^{-\frac{h}{8{,}200}} \sec\psi \qquad [6.48]$$

6.7. Calculation of solar irradiance

6.7.1. *Irradiance on a plane always oriented toward the Sun*

The irradiance of extra-atmospheric solar radiation on a surface perpendicular to the ray is designated by the variable S_{p0} ($S_{p0} = 1{,}367$ W m^{-2}), while the direct solar irradiance on a terrestrial surface perpendicular to the ray is designated by the variable S_{pb}. These two quantities are measured in W m^{-2}. They are linked by a relationship in which atmospheric transmittance is introduced, a dimensionless number whose value for a blue sky is between 0.65 and 0.75, according to Campbell and Norman (1998):

$$S_{pb} = S_{p0}\tau^{m} \qquad [6.49]$$

In equation [6.48], the number m tends toward infinity when the radius becomes horizontal. Regardless of the value of transmittance (which, by definition, is less than one), direct irradiance therefore tends toward zero when the Sun sets on the horizon. A transmittance equal to 1 would correspond to a perfectly transparent medium, that is, a vacuum.

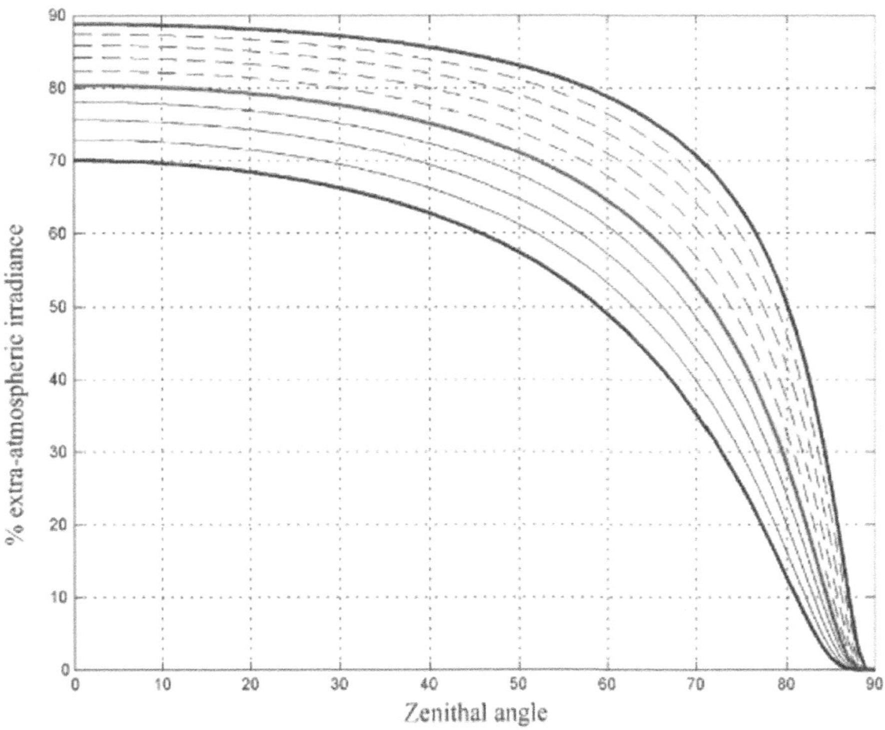

Figure 6.6. *Extra-atmospheric irradiance*

Figure 6.6 shows the percentage S_{pb}/S_{p0} of extra-atmospheric irradiance on a surface perpendicular to the ray as a function of the zenith angle (degrees) for altitudes from 0 to 9,000 m with atmospheric transmittance equal to 0.7 (altitude of 0 m in black, 4,000 m in red, and 9,000 m in blue).

Table 6.2 gives some values for the proportions of extra-atmospheric irradiance.

Proportion of extra-atmospheric irradiance produced by a vertical ray on a horizontal target at different altitudes		
Proportion of extra-atmospheric irradiance	70%	at 0 m altitude
	80%	at 3,850 m
	90%	at 10,000 m
Although without physical justification, we also obtain:	99%	at 30,000 m

Table 6.2. *Proportion of extra-atmospheric irradiance*

At an altitude of 4,000 m, for a vertical radius, we therefore gain 10% of extra-atmospheric radiation. For larger zenith angles, the effect of altitude is more pronounced. Thus, for a point located at 50° north, we obtain the percentages of extra-atmospheric irradiance shown in Table 6.3.

Date	Zenith angle at noon	% at 0 m	% at 4,000 m	Ratio 4,000 m/0 m
June 21	26.5°	76	84	111%
December 21	73.5°	32	49	152%

Table 6.3. *Proportion of extra-atmospheric irradiance at 50° north*

The increase in radiation with altitude is almost the same as for zenithal radiation, 11%, in summer when the zenith angle of the Sun is at its minimum. In winter, however, it is 52%, but at this time of year, the Sun only rises to 16.5° above the horizon.

6.7.2. *Irradiance on a horizontal plane*

Direct irradiance on a horizontal surface is designated by the variable S_b, diffuse irradiance on a horizontal surface by the variable S_d and total irradiance on a horizontal surface by the variable S_t:

$$\begin{aligned} S_b &= S_{pb}\cos\theta = \tau^n S_{p0}\cos\theta \\ S_d &= 0.3\left(1-\tau^n\right)S_{p0}\cos\theta \\ S_t &= S_b + S_d = \left(0.3+0.7\tau^m\right)S_{p0}\cos\theta \end{aligned} \qquad [6.50]$$

Thanks to the contribution of its diffuse component, total irradiance S_t (often referred to as global irradiance) can be greater than direct solar irradiance on a perpendicular surface S_{pb}. When the sun's rays are close to vertical, this situation automatically occurs. This is because some of the energy absorbed by the atmosphere is returned in the form of diffuse energy.

As expected, diffuse contribution decreases with altitude. It can also be observed that at an altitude of 10,000 m, more than 90% of extra-atmospheric irradiance is received. However, questions may be raised about the validity of the model at high altitude (troposphere boundary).

6.7.3. *Analemma*

The *analemma* is the "8" shape traced in the sky by the different positions of the Sun recorded at the same time and from the same place over the course of a calendar year (Figures 6.7(a), 6.8 and 6.9(a)).

To understand the phenomenon, we first need to look at what happens at the poles. We know that the Sun is visible there 24 h a day in summer. It is at its highest on June 21 at the North Pole and on December 21 at the South Pole. If the Earth's trajectory were circular, the analemmas would be identical rays at both poles, varying between the value corresponding to the polar circles and unity. Since the circles in Figure 6.7 are stereographic projections of the celestial sphere (Beckers et al. 2011), the value of the radius corresponding to a height of 27.5° is 0.61.

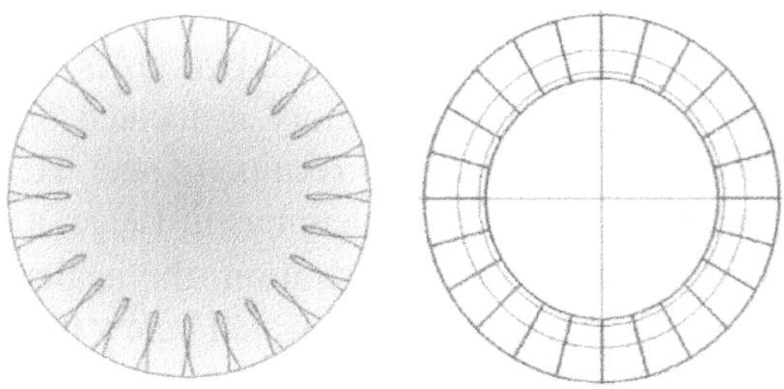

a) b)

Figure 6.7. *North Pole, elliptical orbit (a) and circular orbit (b) (Heliodon)*

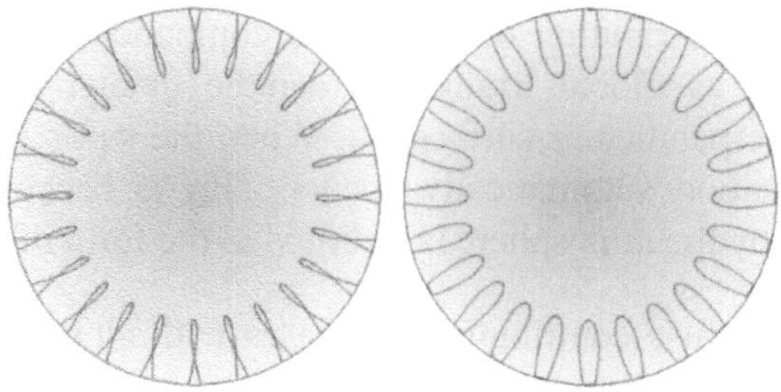

a) b)

Figure 6.8. *Analemmas at both poles (North (a), South (b))*

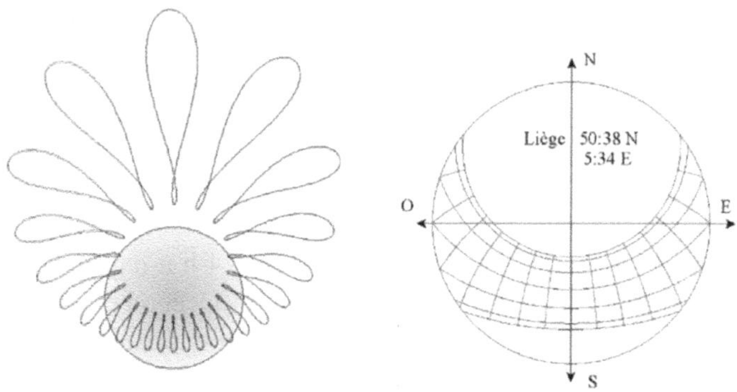

a) b)

Figure 6.9. *Analemma and diagram obtained in Heliodon*

Figure 6.8 compares the analemmas of the two poles. They are different because the perigee of the Earth's trajectory is located on the side of the South Pole. Finally, we compare the results of the Heliodon program (Beckers and Masset 2008) with the analemmas in Liège, a city located at a latitude of almost 51° north. In this illustration, the analemmas have been traced in their entirety, including the parts located below the horizon.

6.7.4. *Energy calculation*

Knowing the length of the day for any latitude and position of the Sun, we can integrate the irradiance over a period of time. This allows us to track the evolution of the energy received throughout the year (Beckers and

Beckers 2014a). Depending on the discipline or convention, energy is expressed in joules (most often MJ) or kWh (1 kWh = 3.6 MJ).

The energies radiated per square meter are time integrals of irradiance, for example, over a day:

$$J_{pb} = \int_{day} S_{pb} dt \qquad [6.51]$$

The purpose of these calculations is to obtain an initial idea of the spatial and temporal distribution of energy from solar irradiance. In the method for calculating solar energy inputs described above, we took into account the actual trajectory of the Earth and the properties of light attenuation by the atmosphere in the case of a clear sky. The calculations can be performed at any point on the globe and at any altitude.

For many applications, however, the idealization of a perfectly circular orbit has the advantage of greatly simplifying representations without changing the calculations, since the elliptical trajectory only causes slight shifts without altering the length of the day.

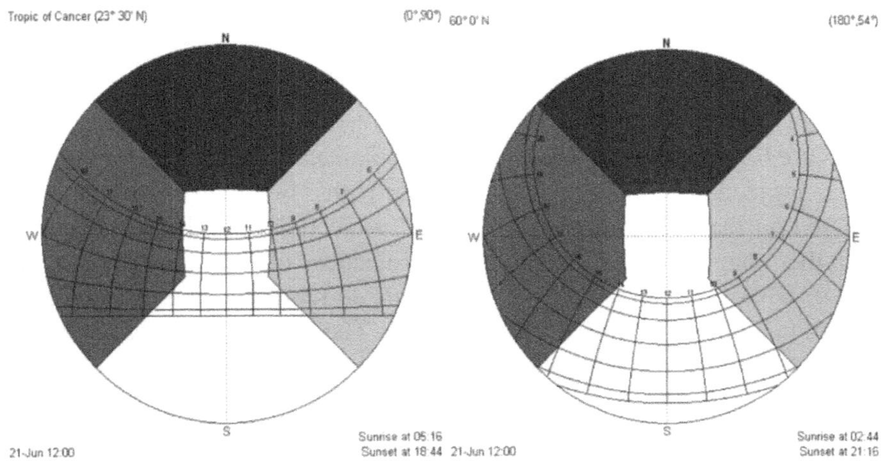

Figure 6.10. *Data at the Tropic of Cancer and at a latitude of 60° north*

To illustrate these concepts, we will use the example of a cubic enclosure with sides measuring 1 m, from which the upper horizontal section and the front face on the south side have been removed. This is a model of a courtyard that is open to the south.

The simulation is performed using Heliodon software (Beckers and Masset 2008). Calculations are performed over the entire day of June 21 (24 h) with an integration step of 5 min. The calculation is performed on the Tropic of Cancer and at 60° north (Stockholm is at 59° 19′ 46″ north and Helsinki at 60° 10′ 15″ north). The solar constant is: $Sp0$ = 1,367 W m^{-2} [6.49]. Atmospheric attenuation is equal to 0.7.

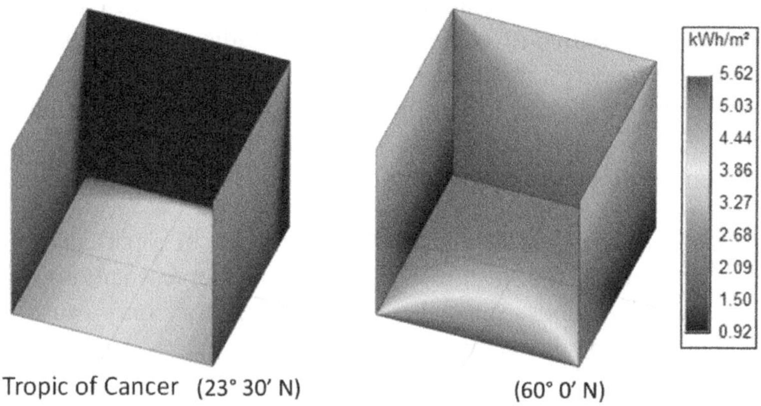

Figure 6.11. *Received energy densities, Tropic of Cancer and latitude 60° north*

Figure 6.10 shows two solar diagrams, obtained in the traditional way, by double projection of the scene and the solar paths: first onto a celestial vault centered on the observation point (central projection onto the sphere), then onto the image plane (stereographic projection of the sphere (Beckers 2011)), while in Figure 6.11, the enclosure is represented in axonometric projection, on June 21 every 5 min for 24 h. The balance of energy received (Figure 6.11) is shown in Table 6.4.

Energy balance	Tropic of Cancer	Stockholm, Helsinki (60° N)
Back wall	0.0 kWh	2.3 kWh
Side walls	1.7 kWh	2.1 kWh
Floor	3.2 kWh	3.0 kWh
Total (four walls)	6.6 kWh	9.4 kWh

Table 6.4. *Energy balance at both latitudes on June 21*

The stereographs are taken from the center of the ground, indicated by a dotted cross on the axonometric projections. They show that the duration of sunshine at this point on June 21 is approximately 4 h at both latitudes. As the Sun passes much higher over the tropics (reaching its zenith at noon), atmospheric attenuation is lower, and the energy density received by the ground is greater than at latitude 60, as shown by the colored maps in the axonometric projections. On the other hand, if we move south of the box, the duration of sunshine remains constant at the tropics, while it increases considerably at latitude 60, because the Sun passes to the south and can penetrate the box obliquely. Overall, however, as shown in Table 6.4, the ground receives slightly more radiation at the tropics, while the side walls receive less. The higher the Sun rises in the sky, the more its radiation strikes the vertical walls at a grazing angle. In fact, the south-facing back wall receives virtually no radiation at the tropics, while it is well exposed at latitude 60.

Figure 6.12 shows two projections of the hemisphere onto which the scene is projected, the first being an equivalent azimuthal projection and the second an orthogonal projection (analogy of Nusselt, 1928).

The first allows us to calculate the amount of sky seen from each point in the scene by evaluating the solid angle. The gray bar shows the proportion of sky (% of the hemisphere area) seen from each point. The gray coloring remains continuous on the edges and tends toward 50% at the top of the three vertical walls. The second shows the sky view factor. The gray bar is the same as for the first graph, but in this case, it is a percentage of the view factor (% of the area of the base disk of the hemisphere). The coloring shows strong discontinuities on the edges of the enclosure.

The solid angle subtended by the sky can be calculated from any point in space. However, the concept of view factor only makes sense for a point belonging to a surface, and its evaluation involves the normal to that surface. Thus, two points that are very close together but located on either side of the edge formed, for example by a horizontal plane and a vertical plane, are associated with normals that are orthogonal to each other and will therefore have very different sky view factors. This explains the discontinuities that appear in the coloring of the figure on the right, which are also found in reality. The preliminary calculation of all the view factors between each pair of surfaces in the scene allows for a rapid evaluation of the infinite number of diffuse light reflections in the scene using the radiosity method.

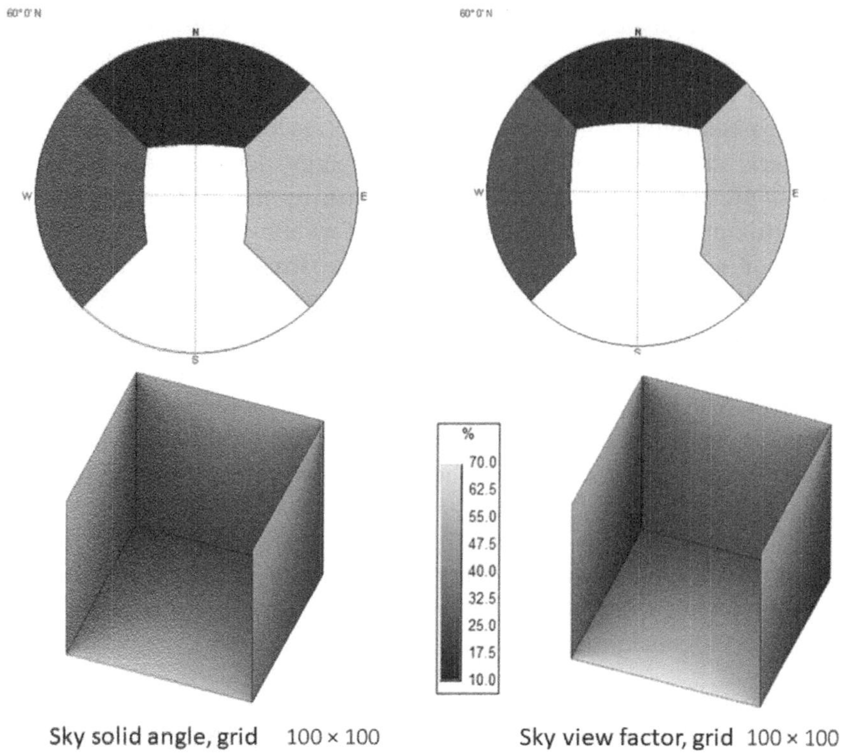

Figure 6.12. *Sky solid angle and sky view factor*

6.8. Radiosity equations

Radiosity equations are used to calculate heat exchange through radiation between elements in the scene (Beckers and Beckers 2012a). They are derived from the definition of radiosity (Beckers and Beckers 2014c):

$$B_i = E_i + \rho_i \sum_{k=1}^{n} B_k F_{ik} \quad [6.52]$$

Radiosity B_i (W m^{-2}) is the total power per unit area leaving patch i. The term E_i denotes the power emitted (exitance) per unit area E (W m^{-2}). The coefficients F_{ik} of matrix $[F]$ represent the view factors from patches i to patches k. For the cube, $[F]$ is given in Table 6.6.

The physical and geometric characteristics of the patches are entered into the two diagonal matrices in Table 6.5. Each term in matrix [R] corresponds to the reflection coefficient of the patch.

Diagonal matrices used in formulations		
Reflectances	R	$R_{ij} = \rho_i \delta_{ij}$
Patch areas	A	$A_{ij} = A_i \delta_{ij}$

Table 6.5. *Diagonal matrices used in solving radiosity equations*

Let I be the unit matrix ($I_{ij} = \delta_{ij}$). In matrix form, the radiosity equation is written as:

$$\left(I - RF\right)B = E \text{ or } MB = E \qquad [6.53]$$

The matrix $M = I - RF$ is the radiosity matrix. Table 6.6 shows, on the left, the view factors for the six faces of the cube already presented in Table 5.3 and, on the right, the radiosity matrix corresponding to a reflection coefficient of 0.5.

View factor matrix	**Radiosity matrix, $\rho = 0.5$**
0.0, 0.2, 0.2, 0.2, 0.2, 0.2, 0.2, 0.0, 0.2, 0.2, 0.2, 0.2, 0.2, 0.2, 0.0, 0.2, 0.2, 0.2, 0.2, 0.2, 0.2, 0.0, 0.2, 0.2, 0.2, 0.2, 0.2, 0.2, 0.0, 0.2, 0.2, 0.2, 0.2, 0.2, 0.2, 0.0	1.0, –0.1, –0.1, –0.1, –0.1, –0.1, –0.1, 1.0, –0.1, –0.1, –0.1, –0.1, –0.1, –0.1, 1.0, –0.1, –0.1, –0.1, –0.1, –0.1, –0.1, 1.0, –0.1, –0.1, –0.1, –0.1, –0.1, –0.1, 1.0, –0.1, –0.1, –0.1, –0.1, –0.1, –0.1, 1.0

Table 6.6. *View factor [F] and radiosity [M = I – RF] matrices*

With $E = [1\ 0\ 0\ 0\ 0\ 0]^T$ W m^{-2}, $[M]$ defined in (Table 6.6) for $\rho = 0.5$, the solution of [6.53] provides the radiosities: $B = [M]^{-1}E$ or $B = [1.0909\ 0.1818\ 0.1818\ 0.1818\ 0.1818\ 0.1818]^T$ W m^{-2} (Figure 6.13(b)).

The radiosities are all identical, except for that of the emitting face; the sum of their products by the areas of the squares to which they are attached is equal to 2 *W*.

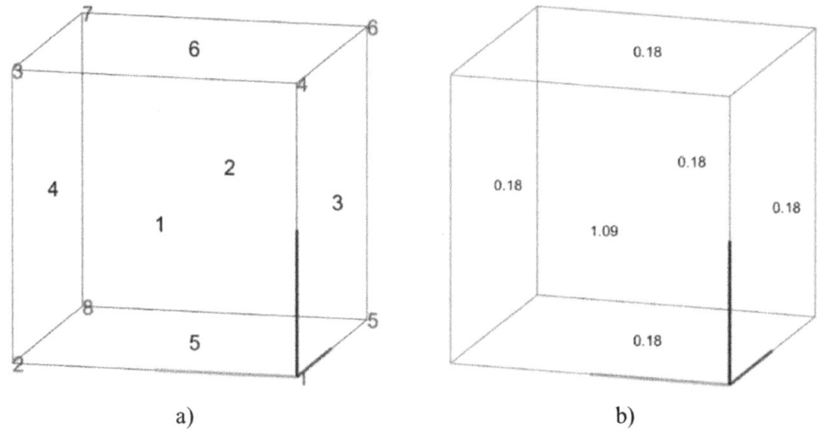

Figure 6.13. *Cube configuration and radiosity for ρ = 0.5*

Figure 6.13(a) shows the numbering of the six faces of the cube. The x, y and z axes are represented by a red, green and blue line segment.

With other values of the reflection coefficient ρ, the results change significantly. For coefficients close to zero, for example, 0.001, they are very small, 0.0002 W m^{-2}. Radiosities increase to significant values when the reflection coefficient tends toward 1 (Table 6.7). For $\rho = 0.999$, the power is 1,000 W; the radiosities are all equal and therefore equal to: 1,000/6 W m^{-2}.

ρ	*B* (1) W m^{-2}	*B* (2 : 6) W m^{-2}	B$_i$ A$_i$, *W*
0.001	1	0.0002	1.001
0.01	1	0.002	1.01
0.4	1.0494	0.1235	1.6667
0.5	1.0909	0.1818	2
0.6	1.1607	0.267	2.5
0.99	17.3623	16.52	100
0.999	167.3612	166.5278	1,000

Table 6.7. *Radiosities based on reflection coefficient*

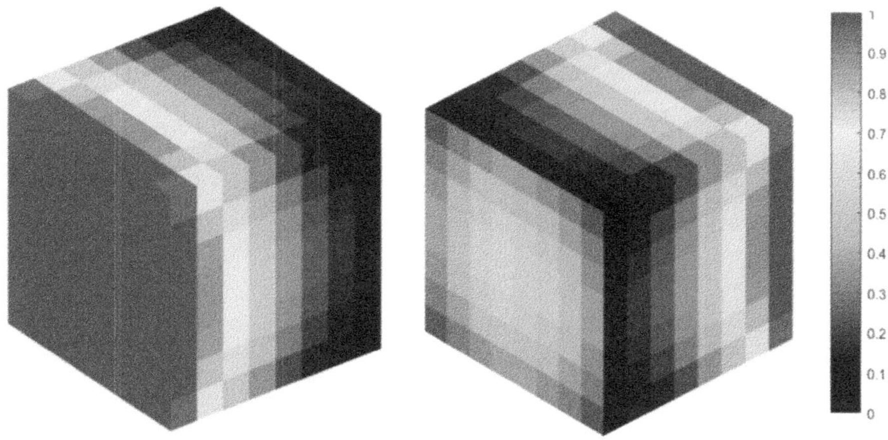

Figure 6.14. *Radiosities on the cube, 8 × 8 mesh on each face*

The view factors for the cube faces given in Table 6.6 correspond to an exact solution. In contrast, the radiosity matrix in the same table corresponds to the coarsest mesh, which has only one element per face. When the same problem is solved with finer meshes, more accurate descriptions of the radiosity field are obtained (Figures 6.14–Figure 6.16).

Figure 6.15. *Radiosities on the cube, 16 × 16 mesh on each face*

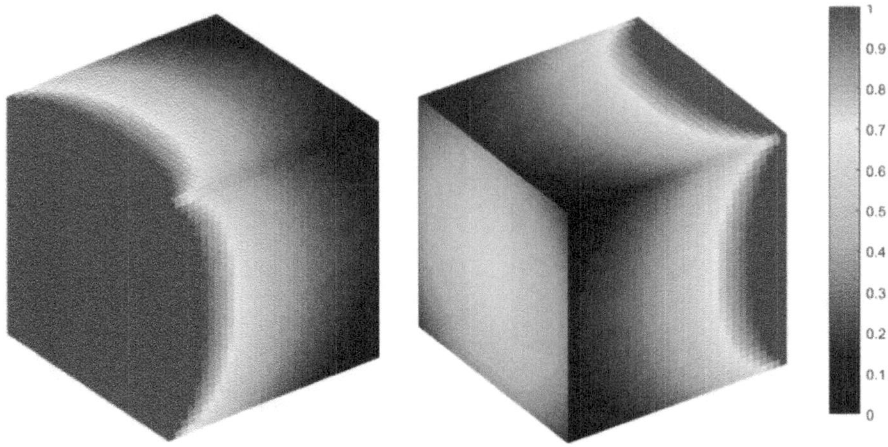

Figure 6.16. *Radiosities on the cube, 32 × 32 mesh on each face*

For the six times 2 × 2 mesh, the maximum and minimum radiosities are 1.0977 W m^{-2} and 0.13293 W m^{-2}. For the mesh of 8 × 8 elements per face, they are 1.1142 W m^{-2} and 0.101 W m^{-2}. For the mesh of 16 × 16 elements per face, they are 1.1195 W m^{-2} and 0.094529 W m^{-2}. Finally, for the mesh of 32 × 32 elements per face (6,144 patches), they are 1.1212 W m^{-2} and 0.090648 W m^{-2}.

Other visualizations of the radiosities distributions on the six faces of the cube are presented in Chapter 8, where, thanks to the use of the Mollweide projection, all six faces are visible simultaneously in the same drawing.

6.9. Extended view factors

The ray tracing method was introduced in Chapter 5 for calculating view factors (Figure 5.9). It is particularly well suited to calculating *extended view factors* (Sillion and Puech 1994; Bugeat et al. 2020). This technique allows the radiosity method to be used for surfaces with specular reflection.

The skin of the area under study behaves like a black body or like a body with a certain degree of reflectivity. This reflectivity can be diffuse or specular. For diffuse reflectivity, Lambert's model is used. Reflectance is independent of the angle of incidence:

$$\rho_r(\theta_i) = \rho_{r0} + (1 - \rho_{r0})\,(1 - \cos\theta_i)^5 \qquad [6.54]$$

	Front	**Back**	**Right**	**Left**	**Botton**	**Top**
Figure 6.13(b) Face	1	2	3	4	5	6
Coordinates	x = 0	x = x1	y = 0	y = y1	z = 0	z = z1
View factor	0.20143	0.19995	0.19812	0.20033	0	0.20017
Extended factor	0.038983	0.038567	0.031533	0	0	0.084883
Bugeat et al. (2020)	0.037	0.037	0.030	0	0	0.078

Table 6.8. *Extended view factors (left face, perfect mirror (Figure 6.13(b)))*

The term $\rho_r(\theta_i)$ refers to the specular reflectance at an angle of incidence θ_i, and $\rho_{r0} = \rho_r$ (0°), the specular reflectance at normal incidence. For simplicity, the value at the normal angle of incidence ρ_{r0} is commonly used to refer to the specular reflectance ρ_r of a material. Bugeat et al. (2020) adopt the simplification of Schlick (1994).

As with the calculation of view factors, each ray represents the same fraction of the view factor ΔF, equal, to within one factor, to the inverse of the number of rays emitted. The view factor *Fij* is therefore equal to the sum of the contributions of the rays emitted from patch *i* and arriving at *j*. The extended part of the view factor comes into play when the ray strikes a surface with specular reflection. The trajectory of the ray is extended in the direction of reflection. This process is repeated until either a perfectly diffuse surface or the sky is reached. The additional part of the extended view factor is then equal to the sum of the weights of the rays originating from patch *i* that intersect patch *j* after one or more specular reflections.

In the cube study, the "*Left*" patch is mirrored. Its specular reflectance is initially equal to one. In this situation, the matrices of the extended view factors and those of the conventional view factors are different. Table 6.8 shows the view factors and the additional part of the extended view factors. The complete view factor matrix is shown in Table 6.9.

The "*Left*" patch does not see any other patches due to specular reflection, as it is the only one with specular reflectance. Its classic and extended view factors are identical.

The "*Right*" patch sees itself by specular reflection through the "*Left*" patch located in front of it. This translates to $F^{extended}_{Right-Right} \neq 0$. The diagonal of the extended view factor matrix may not be zero, unlike that of the conventional view factors.

The sum of the extended view factors of a patch is greater than 1, unlike the sum of the conventional view factors. The closure property, which applies to conventional view factors, no longer applies to extended view factors.

Face	Extended view factor matrix
Bottom	0, 0.200, 0.237, 0.278, 0.237, 0.230
Left	0.200, 0, 0.200, 0.200, 0.200, 0.200
Back	0.237, 0.200, 0, 0.237, 0.278, 0.230
Top	0.278, 0.200, 0.237, 0, 0.237, 0.230
Front	0.237, 0.200, 0.278, 0.237, 0, 0.230
Right	0.230, 0.200, 0.230, 0.230, 0.230, 0.062

Table 6.9. *Extended view factor matrix (left face is a mirror (Figure 6.13(b)))*

7

The Earth and Long Waves

In this chapter, the scene consists of gray bodies. These are solids characterized, in terms of heat diffusion, by their conductivity and thermal capacity. Their surfaces emit radiation, like black bodies, according to the fourth power of their temperature, but this radiation is penalized by an emissivity lower than unity. According to Kirchhoff's law, their absorptivity is therefore also less than unity. This means that part of the radiation that reaches them is reflected or transmitted. Here, we first assume perfectly diffuse emission (black or gray bodies) and perfectly diffuse reflection (gray bodies). The coupling between conduction and radiation is illustrated by the example of a cubic box. The equation that describes the radiation between its inner faces is that of radiosity. The extended view factors introduced in Chapter 5 therefore also apply here and allow us to remove all the restrictive assumptions of the gray body idealization (Lee and Jackson 1976).

7.1. Longwave heat transfer

In shortwave heat transfer, the radiosity equations are decoupled from the diffusion equations. In the case of long waves, these two equations are coupled and must be solved simultaneously. They can still be deduced from a variational principle (Washizu 1955; Breitbach et al. 1990). The balance of radiative heat exchange at the surface of the solid is expressed by writing

For a color version of all the figures in this chapter, see www.iste.co.uk/beckers/finite.zip.

that radiosity B is the sum of exitance E and reflected irradiance $[R]\ J$. The three quantities B, E and J are expressed in W m^{-2}:

$$B = E + [R]J \qquad [7.1]$$

The diagonal matrix $[R]$, such that $R_{ij} = \rho_i\ \delta_{ij}$, contains the reflection coefficients fulfilling Kirchhoff's law [5.18]. In each patch i, it establishes the relationship between emission and reflection when the surface is gray and diffuse. The following dimensionless quantities are used: the absorption coefficient α_i, the emissivity ε_i and the reflectivity ρ_i. The incident irradiance J is related to the other elements of the scene by the view factor matrix $[F]$. It is linked to the sky, when visible, by the sky view factor matrix $[Fc]$ and to the infinite horizontal plane assimilated to the ground, if visible, by the ground view factor matrix $[Fs]$ (Beckers 2013).

$$J = [F]B + [F_c]E_c + [F_s]E_s \qquad [7.2]$$

By substituting the incident irradiance J defined above into [7.1], we obtain:

$$B = E + [R]\ \{[F]B + [F_c]E_c + [F_s]E_s\} \qquad [7.3]$$

From the view factor matrix $[F]$ defined in sections 5.6 and 5.7 and the matrix $[R]$ introduced in [7.1], knowing that matrix $[I]$ is the identity matrix, we obtain the *radiosity matrix* $[M]$:

$$[M] = [I] - [R][F] \qquad [7.4]$$

Exitance E can be expressed as a function of radiosity B:

$$\begin{aligned} E &= B - [R][F]B - [R]\{[F_c]E_c + [F_s]E_s\} \\ &= \{[I] - [R][F]\}B - [R]\{[F_c]E_c + [F_s]E_s\} \\ &= [M]B - [R]\{[F_c]E_c + [F_s]E_s\} \end{aligned} \qquad [7.5]$$

From this relationship, we extract radiosity as a function of exitances:

$$B = [M]^{-1}\left\{E + [R]\left\{F_{sky}E_{sky} + F_{gr}E_{gr}\right\}\right\} \quad [7.6]$$

The radiative load Q (W m^{-2}) is obtained by subtracting the incident irradiance J (W m^{-2}) from the outgoing radiative flux B (W m^{-2}). Q is the heat flux entering or leaving the solid modeled by the finite element mesh:

$$Q = B - J \quad [7.7]$$

Based on [7.1] and [7.2], we obtain:

$$Q = E + \left\{[R] - [I]\right\} J \quad [7.8]$$

The components E_i (W m^{-2}) of the exitance E introduced in [7.5] and explained in [7.8] are calculated based on the components of the surface temperature field T_i, the elements of the boundary i of the solid, their emissivities ε_i and the Stefan–Boltzmann constant σ (W m^{-2} K^{-4}):

$$E \rightarrow E_i = \varepsilon_i \, \sigma \, T_i^4 \quad [7.9]$$

7.2. Three conventional surfaces

– *Mirror*: if the surfaces are perfectly reflective (like mirrors): $[R] = [I]$, the emissivities ε_i are equal to zero, as are the exitances E. As a result, equation [7.4] can no longer be applied, because the matrix $[M]$ is singular. The surface is adiabatic and does not emit or absorb any heat flux. We therefore have: $Q = E = 0$.

– *Black body*: for a black body, $[R] = 0$, the emissivities ε_i are equal to 1, and the exitance E is proportional to the Stefan–Boltzmann constant and the fourth power of the surface temperatures [7.9]. The radiosity matrix [7.5] is an identity matrix: $[M] = [I]$. Radiosity is equal to exitance: $[B] = [E]$, [7.10]. Equation [7.7] becomes:

$$Q = E - J \quad [7.10]$$

– *Gray body*: the first step is to express equation [7.5] in terms of temperatures. In this relationship, the vector E depends on temperatures to the fourth power.

The exitances of the sky and ground are obtained in the same way as in [7.5]. We consider that if $\rho_c = 0$, $\varepsilon_c = 1$ (the same applies to the ground: if $\rho_s = 0$, $\varepsilon_s = 1$):

$$E_c \rightarrow E_c = \varepsilon_c\ \sigma\ T_c^4 \text{ (if emissivity of the sky} = 1, \text{ exitance} = \sigma\ T_c^4) \qquad [7.11]$$

$$E_s \rightarrow E_s = \varepsilon_s\ \sigma\ T_s^4 \text{ (if emissivity of the ground} = 1, \text{ exitance} = \sigma\ T_s^4) \qquad [7.12]$$

In relations [7.7], [7.8] and [7.10], the radiative load Q is a function of the surface temperature of the radiating solid and can therefore be injected into the finite element model:

$$\begin{aligned} &Q = \{[I]-[F]\}[M]^{-1}E(\tau) + \left\{[I]-[F][M]^{-1}\right\}[R] \\ &\left\{[F_c]\sigma T_c^4 + [F_s]\sigma T_s^4\right\} - \left\{[F_c]\sigma T_c^4 + [F_s]\sigma T_s^4\right\} \end{aligned} \qquad [7.13]$$

The last two terms in this expression correspond to radiation related to the sky and the ground, which have imposed temperatures. In the absence of radiation interaction with the sky and the ground, [7.13] simplifies to:

$$Q = \{[I]-[F]\}[M]^{-1}E(\tau) \qquad [7.14]$$

In the last two relations, the exitance $E(\tau)$, calculated in [7.9], changes with each iteration. The term Q is the second classical member of the finite element system. All quantities are expressed in variables at the boundary of the meshed domain, which consists of $3D$ volume elements, with the boundary composed of Coons patches.

In general, there are more nodes than radiating surfaces. For example, a cube has six faces and eight vertices (Figure 6.13). Rules must therefore be established to create a uniform distribution of discretized fields and comply with equilibrium equations (Beckers and Beckers 2012a).

One problem remains to be solved: vector [E] [7.1] concerns surface elements whose temperatures T_i are raised to the fourth power, resulting in a highly nonlinear finite element problem. It is calculated using the temperature value from the previous iteration $n - 1$:

$$E_i = \varepsilon_i \; \sigma \; T_{n-1}^4 = (1\text{-}\rho_i)\sigma \; T_{n-1}^4 \qquad [7.15]$$

The Stefan–Boltzmann constant [5.15] is used: $\sigma = 5.670\,373\ 10^{-8}$ W m^{-2} K^{-4}.

The variable T_{n-1} is the temperature at the center of the surface mesh element at the previous iteration. It can be interpreted as the average temperature of the mesh element skin. For a cavity, if [R] is the diagonal matrix containing the terms ε_i or $(1 - \rho_i)$ from relation [7.1], equation [7.14] becomes:

$$Q = \sigma[R]\{[I]-[F]\}[M]^{-1} \; T_{n-1}^4 \qquad [7.16]$$

The variable Q is expressed in W m^{-2}. To calculate the second member in the finite element system, this radiative load must be multiplied by the area of the patch concerned.

7.3. Cube section with a radiative cavity

We analyze the case of the cavity inside the cube by limiting ourselves to one of the two median vertical planes. Since the problem is symmetrical and the faces parallel to the plane of symmetry are adiabatic, there are no heat fluxes perpendicular to this plane. The problem is therefore two-dimensional. In Figure 7.1, tests are carried out with reflection coefficients equal to 0, 0.5 and 1 over a period of 30 days. The differences are visible on the isotherms and even more so on the representations of the heat flows. For example, we can clearly see the perfect penetration of the black body cavity by the flows and the bypassing of the adiabatic cavity by the flows. Presenting the results in the form of graphic animations helps the user to better understand the physical phenomena in their four dimensions (space and time).

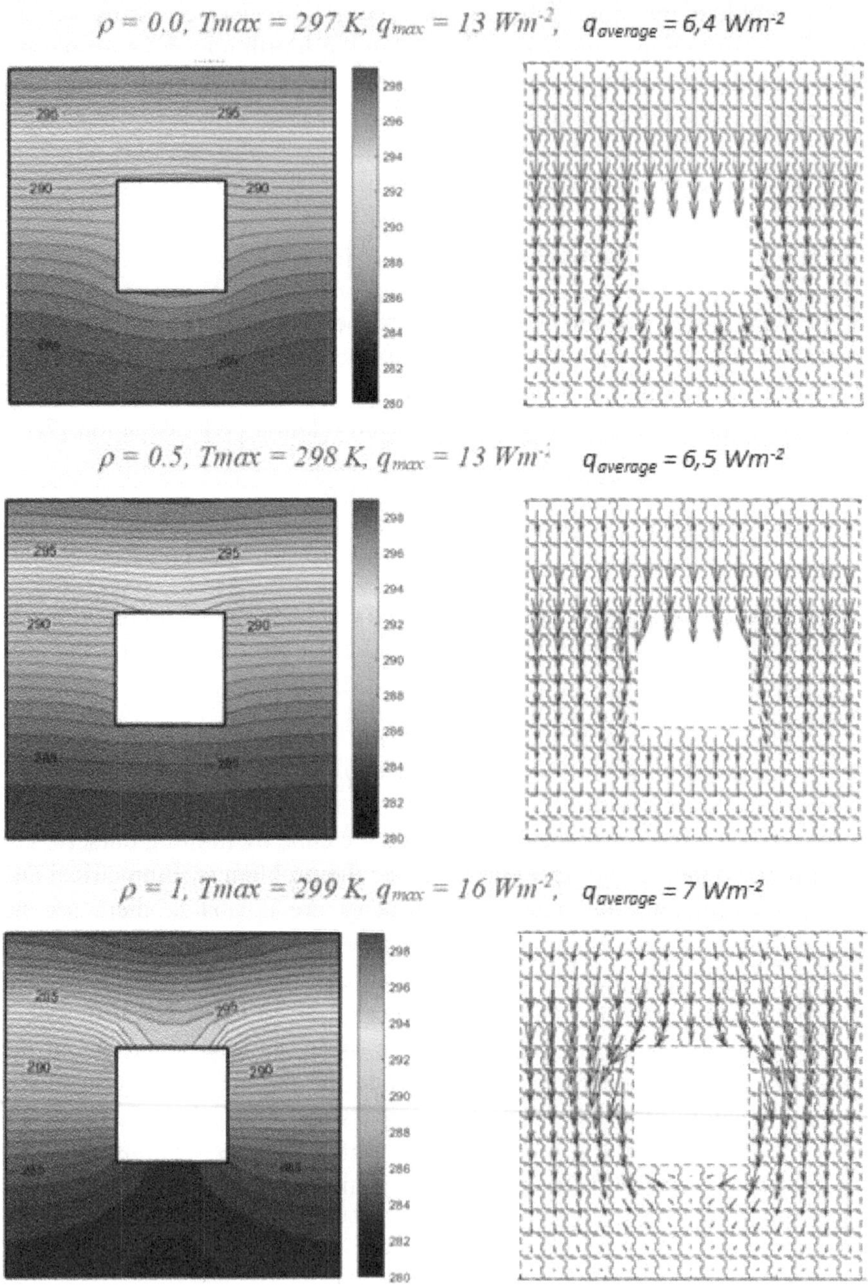

Figure 7.1. *Isotherms and heat flows, black body, gray body, mirror, 30 days*

7.4. Cube with a radiative cavity

For a cube with a single element per face, the relationship between the eight nodal values and the six averages calculated on the patches representing the faces can be written as follows:

$$\begin{bmatrix} T_{f1} \\ T_{f2} \\ T_{f3} \\ T_{f4} \\ T_{f5} \\ T_{f6} \end{bmatrix} = \frac{1}{4} \begin{bmatrix} 1 & 1 & 1 & 1 & 0 & 0 & 0 & 0 \\ 0 & 0 & 0 & 0 & 1 & 1 & 1 & 1 \\ 1 & 0 & 0 & 1 & 1 & 1 & 0 & 0 \\ 0 & 1 & 1 & 0 & 0 & 0 & 1 & 1 \\ 1 & 1 & 0 & 0 & 1 & 0 & 0 & 0 \\ 0 & 0 & 1 & 1 & 0 & 1 & 1 & 0 \end{bmatrix} \begin{bmatrix} T_{n1} \\ T_{n2} \\ T_{n3} \\ T_{n4} \\ T_{n5} \\ T_{n6} \\ T_{n7} \\ T_{n8} \end{bmatrix} \qquad [7.17]$$

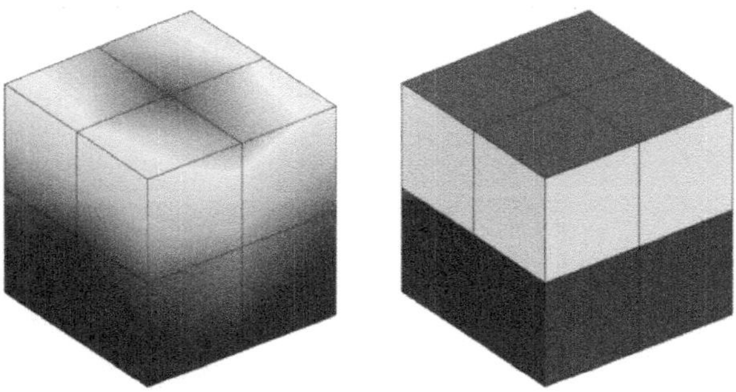

Figure 7.2. *Axonometric projections of the cavity, nodal and average temperatures*

In a unit cube, a cavity is defined whose twelve edges are all 0.64 m long. The surface area of the mesh in the cavity is 2.4576 m^2. For a cavity subject to radiation (Figure 7.2), the dissipation is 3,228 *WK* at the initial stage, 3,295 *WK* at the first iteration and 3,300 *WK* at the second. This value is maintained in subsequent iterations. At the last iteration (no. 3), the temperatures on the cavity patches are, on average, minimum and maximum: 308 K, 304 K and 313 K. The radiative loads Q of the cavity patches are, on average, at minimum, and at maximum: 0 *W*, –1.3933 *W* and 2.2948 *W*.

The sum of the products of the exitances for the average temperature of the patches by the areas of the patches is written as:

$$\left(\sum A_i\right)\sigma\ \varepsilon\ T_m^4 \qquad [7.18]$$

It is equal to 624.7592 *W*. We can also add up the contributions of each patch:

$$\sum A_i Q_i = \sum A_i \sigma\ \varepsilon_i\ T_i^4 \qquad [7.19]$$

In this case, we obtain 625.2 *W*. From this, we can deduce the average radiant temperature of 307.8 K:

$$T_{mr} = \left(\sum A_i Q_i \,/\, \left(\sum A_i \varepsilon_i \sigma\right)\right)^{1/4} \qquad [7.20]$$

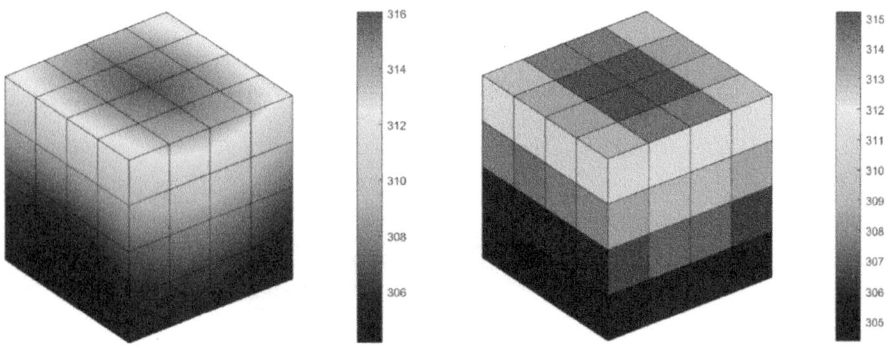

Figure 7.3. *Cavity, nodal and average temperatures (288 hexahedrons)*

With a finer mesh of sixteen elements per face (Figure 7.3), at iteration no. 0, the dissipation is 3,328 *WK*, at iteration no. 1, it is 3,399 *WK*, at no. 2 it is 3,405 *WK*, at iteration no. 3, it is 3,406 *WK* (or 104.2% of the solution with four elements per face). This value is maintained in subsequent iterations. At the final iteration (no. 3), the temperatures on the cavity patches are equal, for the average, minimum and maximum, at 308.2 K, 304.3 K and 315.2 K (or 100.6% of the solution using four elements per face). The radiative loads *Q* of the cavity patches are on average, at minimum and at maximum: 0 *W*, –0.40 *W* and 0.75 *W*. The sum of the products of the exitances corresponding to the average temperatures of the patches by the areas of the patches is equal to 628.8 *W*. Adding the contributions of each patch, we obtain 629.3 *W*. The average radiant temperature [7.20] is estimated at 308.3 K.

As in the case of the terrestrial sphere, all sides of the cavity can be displayed simultaneously on a planisphere (Snyder 1987). For this application, the Mollweide projection will be chosen in the following chapter (Lapaine 2011).

8

Mean Radiant Temperature

In a three-dimensional environment, it is often useful to obtain a compact visualization of what surrounds a point. This is done in two steps. First, everything visible from the point is projected onto the sphere. Next, a planisphere is drawn with distortions compatible with the simulation objectives. The Mollweide projection is *equivalent*: it respects solid angles. If we represent the radiation reaching the point of study from all visible surfaces in the scene, we obtain a kind of "panoramic thermography" of the scene by translating the radiation into surface temperatures, while if we directly add the radiances on the projection, we obtain the mean radiant temperature.

8.1. A cube and two spherical projections

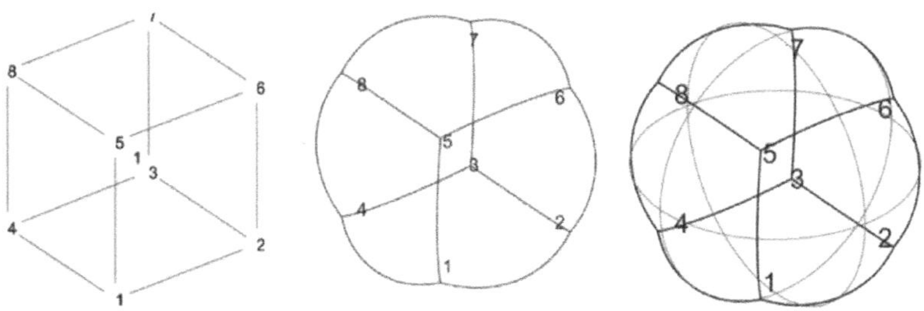

a) b) c)

Figure 8.1. *Axonometric projections of the cube and its projections onto the circumscribed sphere*

For a color version of all the figures in this chapter, see www.iste.co.uk/beckers/finite.zip.

First, we calculate the projection of a cube (Figure 8.1(a)) onto the circumscribed sphere. Its vertices remain unchanged, but its 12 edges become arcs of great circles (Figure 8.1(b)). When using a wireframe representation in which the faces are transparent and only the edges are visible, the entire object can be seen using a single axonometric projection. If the medians of the six faces are added in red (Figure 8.1(c)), the figure is still legible, but it is impossible to include other details because there are too many overlaps. We must therefore consider the faces to be opaque and proceed to eliminate hidden lines.

This type of method is commonly used in CAD and in the finite element method. However, it should be noted that in order to see an object in its entirety, even if it is convex, two axonometric projections must be used, for example, to show the radii on the six faces of the cube (Figures 6.14–6.16, Chapter 6).

8.2. Mollweide projection

To see the entire object in the same illustration, we proceed as we do in mapping the Earth. The terrestrial sphere is commonly represented by a world map, of which there are several types.

Among these possibilities, we select the *Mollweide projection*. This is a pseudo-cylindrical equal-area map projection that displays the surface of a sphere as an ellipse with axes in a ratio of two to one (Beckers and Beckers 2014b). In this projection, the images of the parallels are straight lines. An iterative method is used to calculate the position of the image of a point on the spherical surface on the ellipse of the projection. This projection allows all the faces of a convex volume to be represented on the plane. If this volume is projected onto the sphere, the areas of the spherical polygons representing the faces of the solid are seen in true size in this projection.

The Mollweide projection (Lapaine 2011) of the wireframe representation of the cube in Figure 8.1 is shown in Figure 8.2. The result seems surprising. In fact, only three faces are directly visible: the front face and two vertical faces. The back face appears in two pieces, one on the right and one on the left. The top and bottom faces are only visible from their four sides. Figure 8.2(b) shows all the points of the cube's projection onto the sphere. The boundary of the sphere is represented by the red ellipse, while the 12 edges are shown in black.

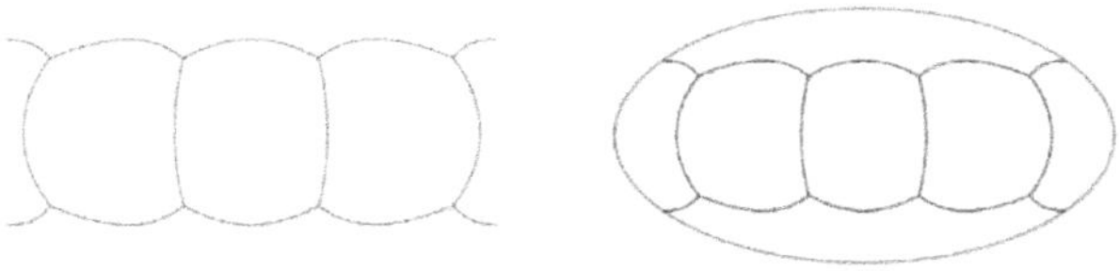

a) b)

Figure 8.2. *Mollweide projection of the 12 edges of the cube*

To make it easier to understand Mollweide's projection, the two medians of the six faces of the cube in Figure 8.1 are added in red. Figure 8.3 is Mollweide's projection of the axonometric projection in Figure 8.1(c). It allows us to verify the equality of the areas of the 24 spherical tiles formed by the spherical projections of the black edges and red medians. Another feature of this projection is that the two outer half-ellipses actually represent the same line extending along the minor axis of the ellipse. The two horizontal faces, therefore, have as their medians the minor axis of the ellipse extended by the two half-ellipses and the branches of the circle centered at the center of the ellipse. In this illustration, the center of the sphere is at the center of the cube.

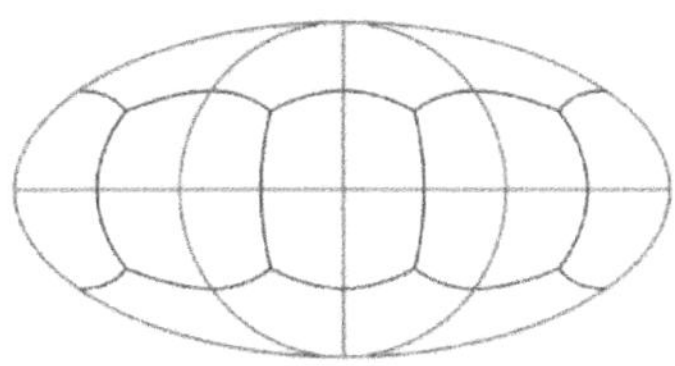

Figure 8.3. *Mollweide projection, 12 edges and 12 medians of the six faces*

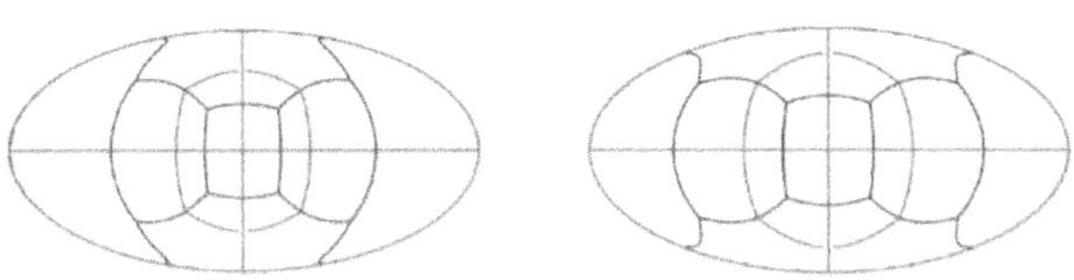

a) b)

Figure 8.4. *Center of the sphere at (a) x = 0.1 m and (b) x = 0.25 m*

Figure 8.4 shows the effect of changing the position of the projection sphere inside the cube.

8.3. Position of a point on the sphere in the Mollweide projection

The A_{ca} area of the spherical cap of a sphere of radius r bounded by the parallel of latitude α (Beckers and Beckers 2014b) is given by the following formula:

$$A_{ca} = 2\pi(1 - \sin\alpha)\, r^2 \qquad [8.1]$$

In a disk with radius b (Figure 8.5), the area A_s of the opening sector 2β is to the area πb^2 of the disk as the opening angle 2β of the sector is to 2π:

$$\frac{A_s}{\pi b^2} = \frac{2\beta}{2\pi} \rightarrow A_s = \beta b^2 \qquad [8.2]$$

The area of the opening sector 2β is therefore equal to βb^2.

The area of the triangle formed by the two radii of the sector and the red intersection line (Figure 8.5(a)) is equal to $b^2 \sin\beta \cos\beta$.

The area of the zone between the chord and the arc of the circle supported by this chord is equal to: $\beta b^2 - b^2 \sin\beta \cos\beta = (\beta - \sin\beta\cos\beta)\, b^2$.

With radius $b = \sqrt{2}$, the result is: $2\beta - 2\sin\beta\cos\beta = 2\beta - \sin 2\beta$.

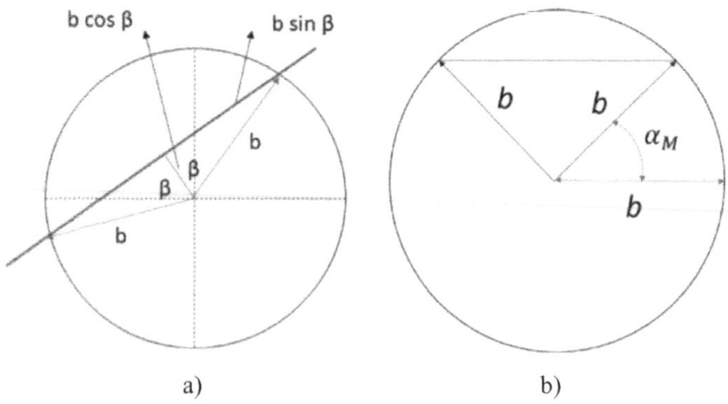

Figure 8.5. *Subdomains separated by a line and area under the upper arc*

By rotating Figure 8.5(a) so that the red secant is horizontal and replacing angle β with its complement: $\alpha_M = \pi/2 - \beta$, the area A_{cc} of the region between the chord and the arc of the circle supported by this chord $2\beta - \sin 2\beta$ can be written as:

$$A_{cc} = \pi - 2\alpha_M - \sin 2\alpha_M \quad [8.3]$$
$$\text{with: } \sin 2\beta = \sin 2\alpha_M$$

To convert to the Mollweide representation, we multiply the x-coordinates and the area between the chord and the ellipse by two.

We can therefore write the equality between twice the area [8.3] and that of the spherical cap [8.1]:

$$2\pi - 4\alpha_M - 2\sin 2\alpha_M = 2\pi - 2\pi\sin a \quad [8.4]$$

Simplifying:

$$2\alpha_M + \sin 2\alpha_M = \pi\sin a \quad [8.5]$$

To calculate the image of a point on the sphere on the Mollweide projection, the first step is therefore to calculate the angle α_M by solving equation [8.5]. This calculation is performed using the Newton–Raphson method. The angle increment α_M, which we denote α_M', is given by:

$$\Delta\alpha_M' = \alpha_M' - (\alpha_M' + \sin\alpha_M' - \pi\sin\alpha) / (1 + \cos\alpha_M') \quad [8.6]$$

As in the reference (Snyder 1987), let us take the latitude of 50° as the initial value, that is α_{M0} = 50 π/180° = 0.8727 radians. The first increment is therefore:

$$\alpha_M^1 = \alpha_M^0 - \left(\alpha_M^0 + \sin\alpha_M^0 - \pi\sin\alpha_M^0\right) / \left(1 + \cos\alpha_M^0\right) \quad [8.7]$$

The first update is:

$$\alpha_M = \alpha_M^n \quad [8.8]$$

For the example in Snyder (1987), where we want to calculate the position of the image of a point with a latitude of –50°, we obtain the latitude of its Mollweide projection $a_M = -40.6289°$. We then define the coordinates (x, y) of the projection of the point on the sphere:

$$\begin{aligned} y &= \sqrt{2}\ r\ \sin a_M \\ x &= \frac{\varphi}{\pi} 2\sqrt{2} r \cos a_M \end{aligned} \qquad [8.9]$$

8.4. Visualization of shortwave radiosities in the cube

To improve the visualization of the radiosity calculation results on the six faces of a cube (Figures 6.14–6.16, section 6.8), Mollweide projections are used to visualize them simultaneously on all six faces (Figure 8.6). To simplify the drawings, the edges of the mesh elements are represented by straight line segments, which is very apparent in the upper left box. The four results correspond to 2 × 2, 4 × 4, 8 × 8 and 16 × 16 meshes. In the latter case, the cube has 1,536 elements.

All patches have the same area on the cube, namely, 0.25 m^2, 0.0625 m^2, 0.0156 m^2, 0.0039 m^2. However, it should be noted that the areas of their spherical projections are not equal (except in the 2 × 2 grid of faces). In the Mollweide projection, the areas of the patches are also different. In all examples, the power, which is the sum of the products of the elementary areas and the radiosities, is equal to 2 *W*.

In the examples in Figures 6.14–6.16 in Chapter 6 and Figures 8.6 and 8.7, the radiances are calculated for a reflectance coefficient of 0.5 and an exitance of 1 W m^{-2} on the front face. In Figure 8.6, the two lower meshes have 384 and 1,536 elements, respectively.

For Figure 8.7, a mesh of 32 × 32 × 6 = 6,144 elements is used. The minimum, maximum and average values of B_i radiance are 0.0906 W m^{-2}, 1.1212 W m^{-2} and 0.3333 W m^{-2}.

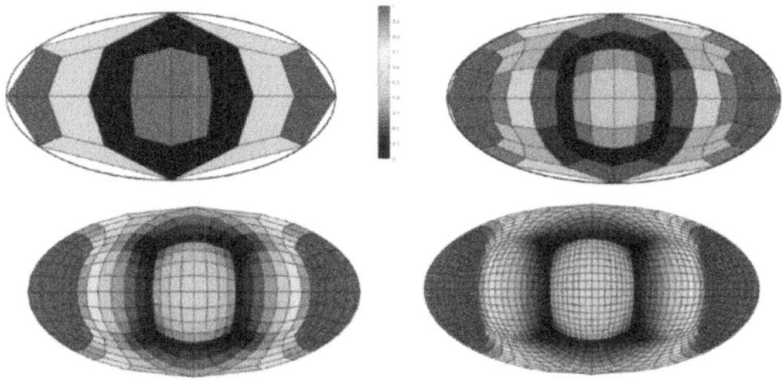

Figure 8.6. *Radiosity maps for four cube face meshes*

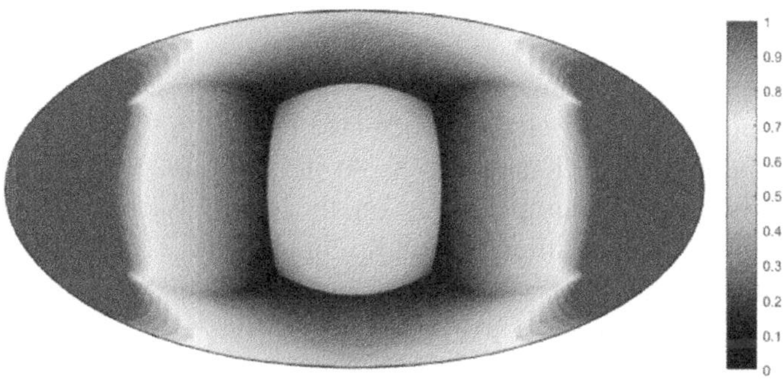

Figure 8.7. *Radiosities in the cube, ρ = 0.5, 6,144 elements*

8.5. Longwave radiation in a cube

In the example shown in Figure 8.8, we can see the average temperatures on the radiative inner faces of a cubic container. The volume of the cavity is 0.26214 m^3 and the length of the edges is 0.64 m.

For the simplified model, the temperatures obtained for the two virtual convection nodes are 300 K and 327.1839 K. At iteration no. 3, the dissipation is 3,300 *WK*. The average temperatures on the cavity skin are, in order of "average, minimum, maximum": 307.7195 K, 304.147 K and

313.2114 K. At iteration no. 3, the radiative loads are, in the same order: 0 *W*, –1.3933 *W* and 2.2948 *W*, and the densities are: 0 W m^{-2}, -13.6062 W m^{-2} and 22.4098 W m^{-2}.

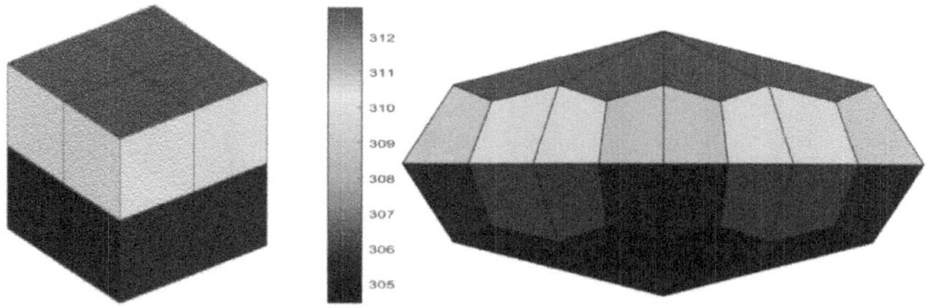

Figure 8.8. *Average temperatures of a 24-element radiative cavity*

For the model with 16 elements per face (Figure 8.9), the virtual nodes have temperatures of 300 K and 328 K. At iteration no. 3, the dissipation is 3406 *WK*. The average temperatures on the cavity skin are, in order of "average, minimum, maximum": 308 K, 304 K and 315 K. The radiative loads are, in the same order: 0 *W*, –0.40 *W* and 0.75 *W*. Their densities are: 0 W m^{-2}, –15.7 W m^{-2} and 29.4 W m^{-2}.

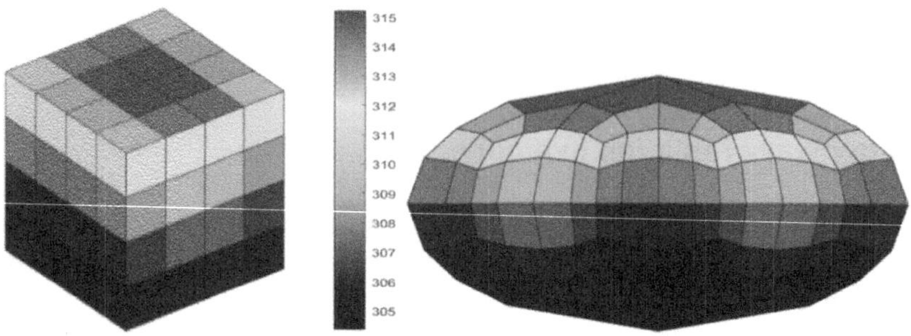

Figure 8.9. *Average temperatures of a radiative cavity with 96 elements*

The radiative loads on each element of the cavity skin are visualized in the same way as the average temperatures (Figure 8.10). These loads are positive on the upper horizontal face and on the tops of some vertical faces, namely on 28 of the 96 elements.

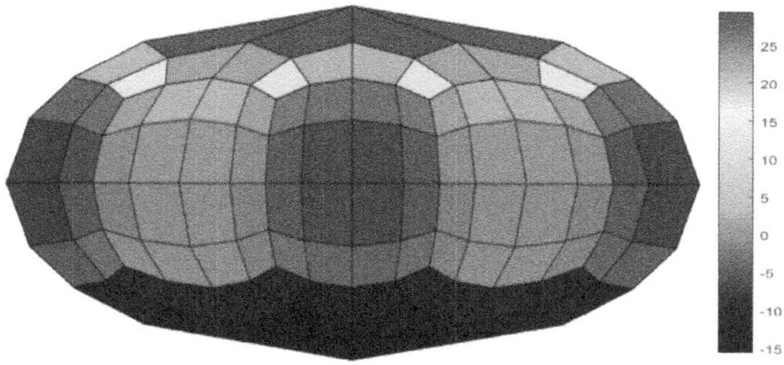

Figure 8.10. *Radiative loads in the cavity (W m^{-2}) at convergence*

8.6. Mollweide projections of the cube and cells of equal area

In this section, Mollweide projections of the cube, possibly meshed, and a number of cells of equal area are superimposed on the same drawing (Figure 8.11).

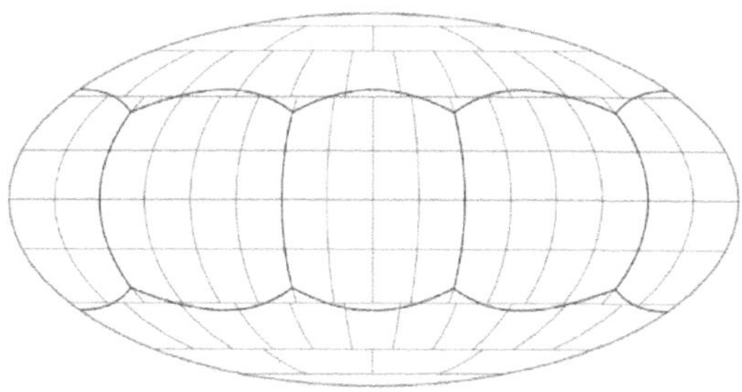

Figure 8.11. *A cube and 96 cells of equal area*

The cube divided into 16 elements per face is shown in Figure 8.12, in axonometric projection on the left and in Mollweide projection on the right. In this example, the sphere is divided into 96 cells (blue). The Mollweide projection of a cube with the same center as the sphere is superimposed on it (black). The sequence of the number of cells per latitude, equal according to the standard procedure to S = [1 5 16 31 48 65 80 91 95 96], is replaced by the sequence S = [1 5 16 32 48 64 80 91 95 96], where the changes are indicated in red, so as to have 16 cells in the four central bands. Since the vertical edges of the cube have spherical projections coinciding with regularly spaced meridians and the cell boundaries are arcs of parallels and meridians, it suffices for the number of cells to be a multiple of four to obtain the superimposition of the cell boundaries and vertical faces of the cube.

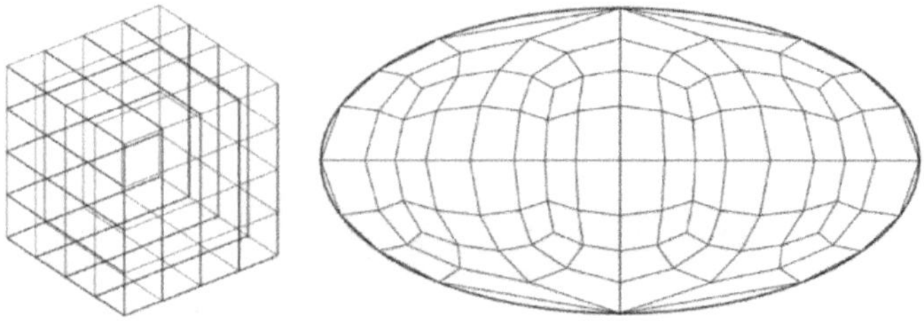

Figure 8.12. *Wireframe and Mollweide axonometry of the cube, 16 elements per face*

A cube whose six faces are divided into 64 square elements is shown in Figure 8.13. On the cube, all these elements have the same area, but this is not the case for their spherical projections and, consequently, for their Mollweide projection. The figure shows the 384 squares created on the faces of the cube.

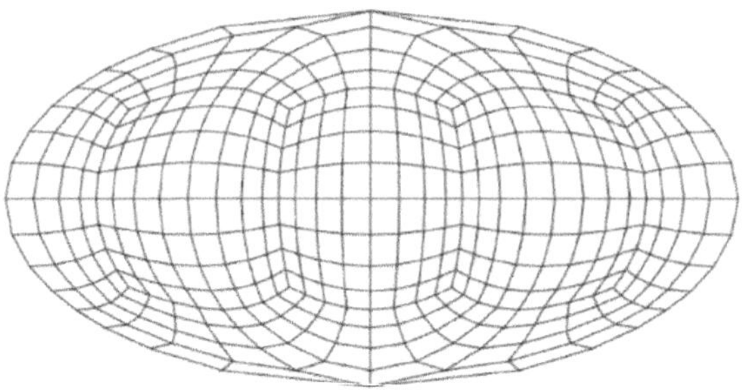

Figure 8.13. *Cube whose faces are subdivided into 64 elements*

By imposing 600 cells in the background mesh, it is easy to verify compliance with the areas (Figure 8.14). The projections of the six faces must each include 100 cells. To facilitate counting, we imposed the presence of 40 cells in the 10 layers around the equator. Since the spherical projections of the sides of the cube are arcs of great circles, the vertical edges are projected along meridians and coincide with the boundaries of certain cells.

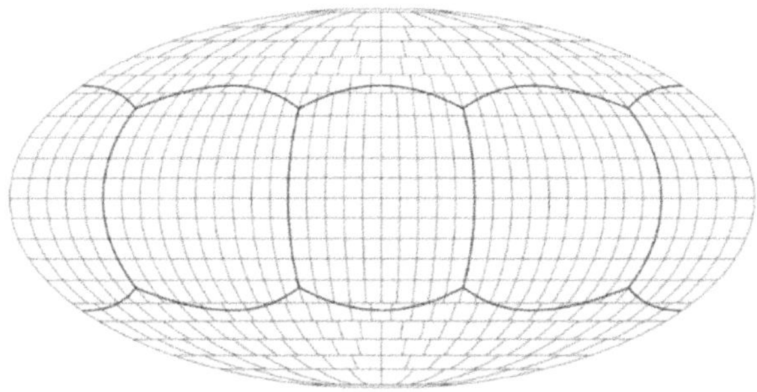

Figure 8.14. *Cube and 600 cells of equal area*

8.7. Use of the ray tracing technique

The grid of the sphere and the rays emitted randomly from each of the cells are shown with the cube in the background (in black) (Figure 8.15). The rays emitted by the cells of the grid appear in the axonometric projection in Figure 8.16. This type of model will be used to calculate the mean radiant temperature.

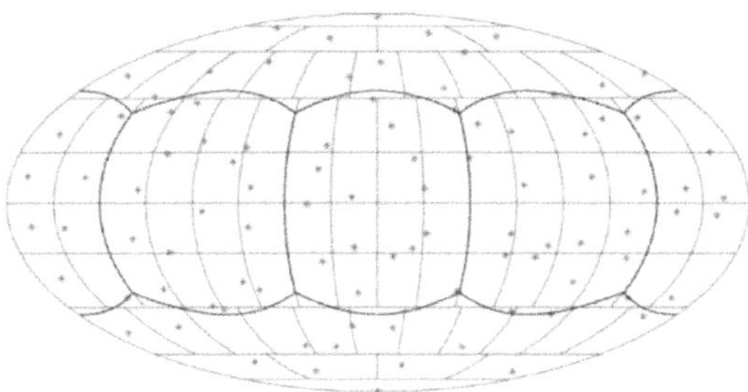

Figure 8.15. *96 cells, 96 rays, cubic cavity*

Instead of visualizing the results on the finite element mesh, a black sphere is defined on which the radiation is calculated at any point in the cavity. The sphere is divided into n cells of equal solid angle. Ray tracing is then performed from the center of the sphere to random points in each of the cells. In Figure 8.15, the cells are shown in blue, the positions of the rays in red and the edges of the cavity in black. The rays and the cavity are also shown in axonometric projection in Figure 8.16. Statistically, the six faces of the cavity should receive 16 rays. In this test, this is indeed the case.

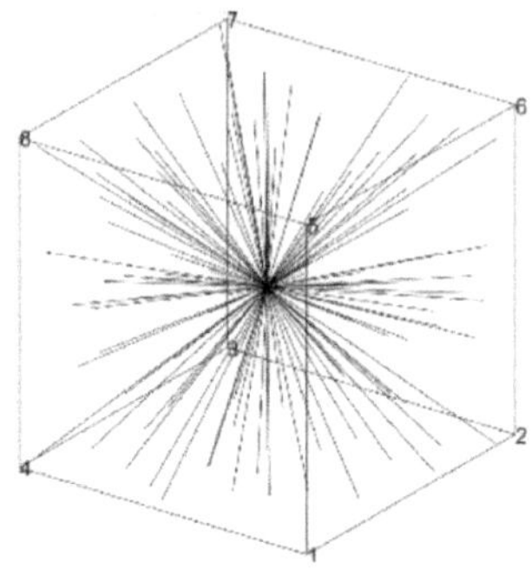

Figure 8.16. *Cavity with 96 rays*

In Figure 8.17, the rays that reach one of the vertical faces and the lower horizontal face are shown in axonometric projection.

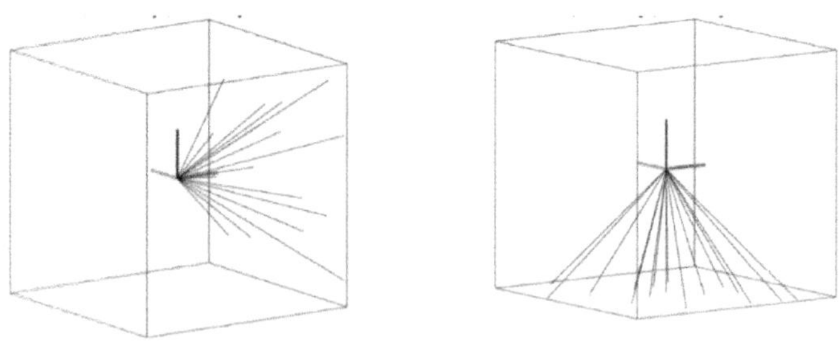

Figure 8.17. *Cavity comprising two groups of rays*

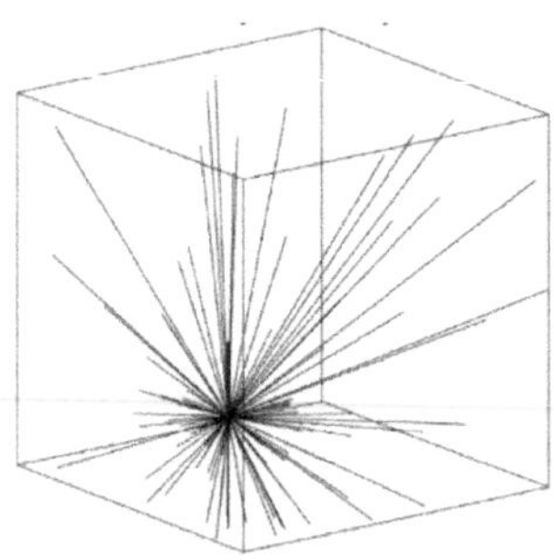

Figure 8.18. *96 rays traced from point*
x = 0.1 m, y = 0.2 m and z = 0.3 m

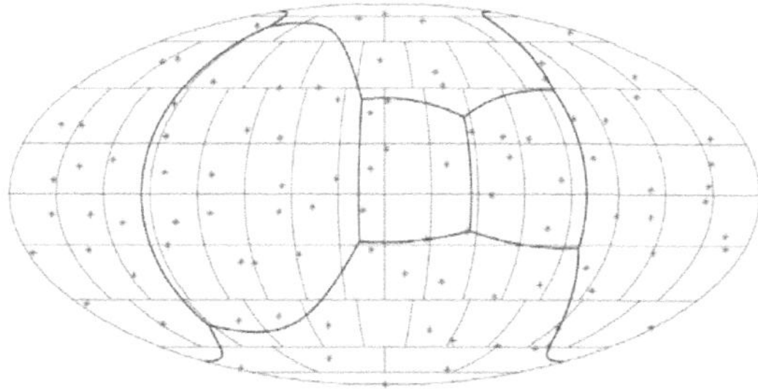

Figure 8.19. *Mollweide from a point not located at the center of the cube*

In the example shown in Figures 8.18 and 8.19, the black sphere is no longer placed at the center of the unit cube, but at the point $x = 0.1$ m, $y = 0.2$ m and $z = 0.3$ m. The distribution of the 96 rays reaching the six faces is: 8, 35, 6, 22, 7 and 18.

For each ray that actually hits one of the cube's faces, it is necessary to determine which element of the mesh has been struck. In the case of a finite element mesh with four elements per face (Figure 8.20(a)), the contours of the 24 faces are shown in Figure 8.20(b) for the center of ray tracing located at [0.9 0.5 0.5].

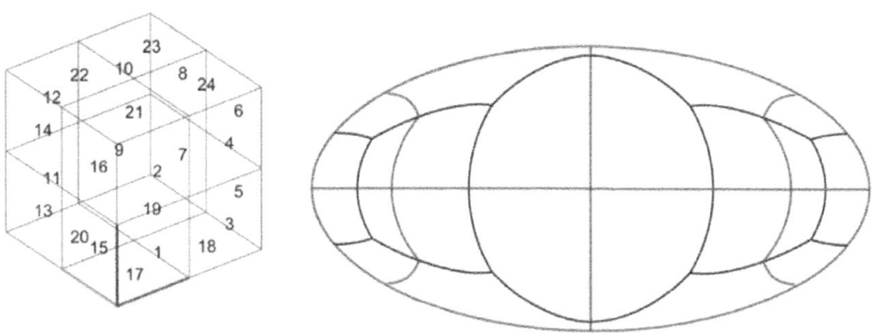

a) b)

Figure 8.20. *Four elements per side, labels, and Mollweide projection from [0.9 0.5 0.5]*

In the example shown in Figure 8.21, the rays are cast from a point located at [0.9 0.5 0.5] m. The Mollweide projection of the rays is shown in Figure 8.22. The edges of the cube are shown in black.

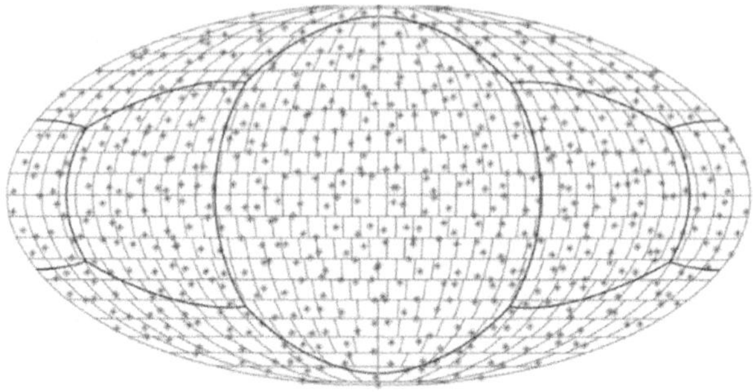

Figure 8.21. *Ray casting of 600 rays from a point located at [0.9 0.5 0.5] m*

The average temperature is 307.3 K. Other results from ray casting of 600 rays are given in Table 8.1.

The second column gives the number of rays intersecting each of the six faces. The third column provides details of the intersections with the four finite elements composing the faces.

The average number of intersections with the elements is 25, with a minimum of 6 and a maximum of 63.

Number of elements	Number of intersections with the cube	
	Faces	Finite elements
5 – 8 (x = 1)	251	63, 61, 64, 63
13 – 16 (x = 0)	43	11, 11, 9, 12
9 – 12 (x = 1)	78	33, 31, 6, 8
1 – 4 (x = 0)	79	34, 33, 6, 6
21 – 24 (x = 1)	75	32, 31, 6, 6
17 – 20 (x = 0)	74	32, 30, 6, 6

Table 8.1. *Points where rays intersect with faces and finite elements*

For a casting of 600 rays, emitted from the center of the cube, the average number of rays reaching each of the 24 elements is 25, the minimum number is 22 and the maximum is 27 (Figure 8.22).

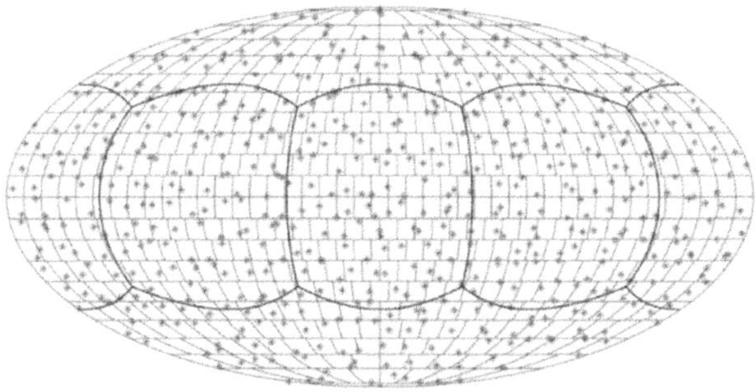

Figure 8.22. *A total of 600 cells and rays emitted from the center of a cube*

8.8. Cube with five cold sides and one hot side

The cube has five faces at 275 K and the sixth at 335 K. The average surface temperature is therefore 285 K.

In the illustration in Figure 8.23, half of the rays intersect the nearest surface, as in Figure 8.24.

The *mean radiant temperature* (T_{MRT}) is obtained by first evaluating the weighted average of the number of rays reaching each face of their irradiance and then calculating the temperature producing this irradiance.

Table 8.2 shows how the T_{MRT} increases as we approach the hot side. It is compared to the average temperature of the surfaces weighted by the solid angle that they subtend from each point considered.

This simplified assessment is, as expected, consistently lower than the correct value of the T_{MRT}.

This example illustrates the fundamental property of thermal calculation incorporating conduction and radiation. At the end of this calculation, a continuous temperature field is obtained, both in the solid and in the surrounding medium, whatever it may be.

Coord. *x*	Coord. *y*	Coord. *z*	No. of rays	Average T, K	Weighted average T, K	T_{MRT}, K
0.001	0.5	0.5	600	285	279.0	280.4
0.1	0.5	0.5	150	285	279.8	281.4
0.25	0.5	0.5	150	285	280.6	282.4
0.5	0.5	0.5	150	285	284.6	287.4
0.5	0.5	0.5	600	285	284.7	287.5
0.75	0.5	0.5	150	285	291.8	295.7
0.75	0.5	0.5	600	285	292.8	296.8
0.90	0.5	0.5	600	285	300.1	304.5
0.99	0.5	0.5	150	285	304.2	308.6
0.999	0.5	0.5	600	285	304.6	309.0

Table 8.2. *Ray tracing results*

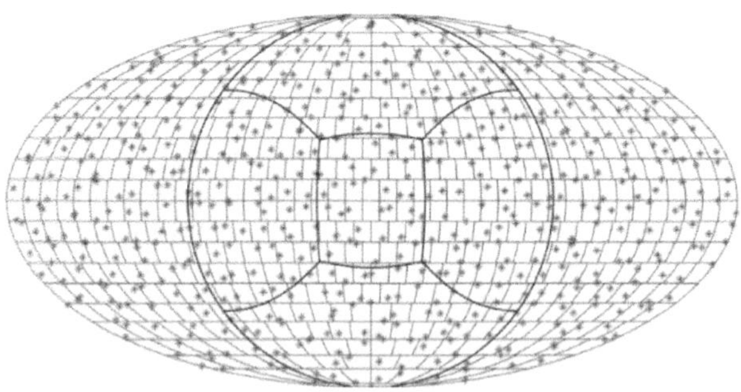

Figure 8.23. *A total of 600 rays emitted from a point located at x = 0.001 m*

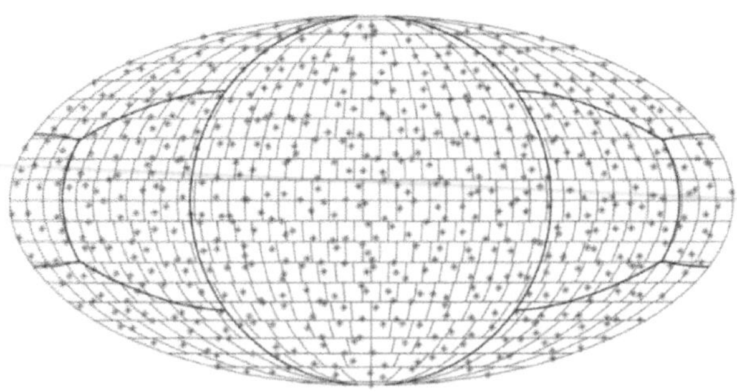

Figure 8.24. *A total of 600 rays emitted from a point located at x = 0.999 m*

Conclusion

The equations of thermal diffusion and their boundary conditions – Dirichlet and Neumann conditions, convection and radiation conditions – are derived in steady state and transient state from a variational principle. Instead of discretizing the partial differential equations, this operation is performed on the minimization conditions of the functional. To do this, a Rayleigh–Ritz decomposition of the functional is performed, followed by a partitioning of the domain into subdomains called finite elements. These must satisfy certain continuity conditions. The functional is thus transformed into an algebraic function whose minimization leads to a system of linear equations. Effective standard methods are available to solve this system. An essential property of the finite element method is its convergence to the exact solution when the mesh is refined.

Problems involving heat exchange through radiation can be simplified by expressing them in the form of radiosity equations. These equations relate the heat fluxes present on the boundary surfaces of the solid. Two situations are analyzed: short waves, corresponding to visible light, and long waves, corresponding to thermal emissions. These equations correspond to diffuse reflection, but they can be generalized using the concept of extended view factor. The variables in this problem are discrete quantities related to the outer faces of the elements constituting the finite element mesh. The system of radiosity equations is linear and can easily be coupled with that of finite elements.

By calculating mean radiant temperatures, it is possible to obtain a temperature field produced by conduction or radiation throughout the entire space. Planisphere-type projection techniques can also be used to represent thermograms at any point in space. In solids, cross-sections or surface temperatures are used. When a cross-section is carried out, it is possible to visualize the heat flows within the cross-section.

The method proposed in this book is best suited for calculating temperature fields and thermographic representations. It simply requires an understanding of the finite element method and the concept of view factor.

References

Appel, A. (1968). Some techniques for shading machine renderings of solids. In *Proceedings of the Spring Joint Computer Conference*. AFIPS, Atlantic City.

Beckers, B. (2009). Geometrical interpretation of sky light in architecture projects. In *Proceedings of the Conférence Internationale Scientifique pour le BATiment*. CISBAT, Lausanne.

Beckers, B. (2011). Impact of solar energy on cities sustainability. In *Proceedings of the 27th Conference on Passive and Low Energy Architecture PLEA*, Louvain-la-Neuve.

Beckers, B. (ed.) (2012). Worldwide aspects of solar radiation impact. In *Solar Energy at Urban Scale*. ISTE Ltd, London and Wiley, Hoboken, NJ.

Beckers, B. (2013). Taking advantage of low radiative coupling in 3D urban models. In *Proceedings of the Eurographics Workshop on Urban Data Modelling and Visualization*. UDMV, Girona.

Beckers, B. (2016). Multiscale analysis as a central component of urban physics modeling. In *Computational Methods for Solids and Fluids, Multiscale Analysis, Probability Aspects and Model Reduction*, Ibrahimbegovic, A. (ed.). Springer, Berlin.

Beckers, B. (2017). Géométrie assistée par ordinateur [Online]. Available at: http://www.heliodon.net/downloads/Beckers%2020180824%20G%C3%A9om%C3%A9trie%20assist%C3%A9e%20par%20ordinateur.pdf [Accessed 15 May 2025].

Beckers, B. (2019). Omnidirectional source modeling for ray tracing in room acoustics with specular reflection. *WSEAS Transactions on Acoustics and Music*, 6, 13–22.

Beckers, B. (2023). Fiammetta, finite element method applied to heat transfer [Online]. Available at: http://www.heliodon.net/downloads/Beckers%2020230830%20-%20Fiammetta,%20Finite%20Element%20Method%20Applied%20to%20Heat%20Transfer.pdf [Accessed 15 May 2025].

Beckers, B. and Beckers, P. (2010a). Les anomalies, l'équation de Kepler, la position du soleil [Online]. Available at: http://www.heliodon.net/downloads/Beckers_2010_Helio_006_fr_2.pdf [Accessed 15 May 2025].

Beckers, B. and Beckers, P. (2010b). Comment calculer la déclinaison du soleil [Online]. Available at: http://www.heliodon.net/downloads/Beckers_2010_Helio_007_fr.pdf [Accessed 15 May 2025].

Beckers, B. and Beckers, P. (2011). Calcul du rayonnement solaire atténué par l'atmosphère [Online]. Available at: http://www.heliodon.net/downloads/Beckers_2011_Helio_008_-_Calcul_du_rayonnement_solaire_attenue_par_latmosphere.pdf [Accessed 15 May 2025].

Beckers, B. and Beckers, P. (2012a). Radiative simulation methods. In *Solar Energy at Urban Scale*, Beckers, B. (ed.). ISTE Ltd, London and Wiley, Hoboken, NJ.

Beckers, B. and Beckers, P. (2012b). A general rule for disk and hemisphere partition into equal-area cells. *Computational Geometry – Theory and Applications*, 45(7), 275–283.

Beckers, B. and Beckers, P. (2014a). Sky vault partition for computing daylight availability and shortwave energy budget on an urban scale. *Lighting Research and Technology*, 46(6), 716–728.

Beckers, B. and Beckers, P. (2014b). *Reconciliation of Geometry and Perception in Radiation Physics*. ISTE Ltd, London and Wiley, Hoboken, NJ.

Beckers, B. and Beckers, P. (2014c). Super element technique for solar energy optimization at urban level. In *Proceedings of An International Conference on Engineering and Applied Sciences Optimization*. OPT-I, Kos.

Beckers, B. and Beckers, P. (2016a). A 33-line heat transfer finite element code [Online]. Available at: http://www.heliodon.net/downloads/Beckers_&_Beckers_2015_Supporting_Information.pdf [Accessed 15 May 2025].

Beckers, B. and Beckers, P. (2016b). Fast and accurate view factor generation. In *Proceedings of An International Conference on Urban Physics*. Ficup, Quito.

Beckers, B. and Beckers, P. (2016c). Complete set of Matlab procedures for achieving uniform ray generation [Online]. Available at: http://www.heliodon.net/downloads/Beckers_&_Beckers%2020180423%20Complete%20set%20of%20Matlab%20procedures%20for%20achieving%20uniform%20ray%20generation.pdf [Accessed 15 May 2025].

Beckers, B. and Beckers, P. (2024). Le corps noir [Online]. Available at: http://www.heliodon.net/downloads/Beckers%2020240426%20-%20Le%20corps%20noir.pdf [Accessed 15 May 2025].

Beckers, B. and Masset, L. (2008). Heliodon2 [Online]. Available at: http://www.heliodon.net/downloads/Heliodon2_User_Guide_2008_fr_11_02.pdf [Accessed 15 May 2025].

Beckers, B., Masset, L., Beckers, P. (2009). Commentaires sur l'analogie de Nusselt [Online]. Available at: http://www.heliodon.net/downloads/Beckers%202009%20Helio_002_fr%20-%20Commentaires%20-%20Analogie%20de%20%20Nusselt.pdf [Accessed 15 May 2025].

Beckers, B., Masset, L., Beckers, P. (2011). The universal projection for computing data carried on the hemisphere. *Computer-Aided Design*, 43(2), 219–226.

Breitbach, G., Altes, J., Sczimarowsky, M. (1990). Solution of radiative problems using variational based finite element method. *International Journal of Numerical Methods in Engineering*, 29, 1701–1714.

Bugeat, A., Beckers, B., Fernandez, E. (2020). Improving the daylighting performance of residential light wells by reflecting and redirecting approaches. *Solar Energy*, 207, 1434–1444.

Campbell, G.S. and Norman, J.M. (1998). *An Introduction to Environmental Biophysics*. Springer, New York.

Coons, S.A. (1967). Surfaces for computer-aided design of space forms. Project MAC-TR-41, Massachusetts Institute of Technology, Cambridge, MA.

Coons, S.A. (1977). *Méthode matricielle. Mathématiques et CAO.* Hermès, Paris.

Courant, R. (1943). Variational methods for solution of problems of equilibrium and vibrations. *Bull. Amer. Math. Soc.*, 49(1), 1–23.

Ergatoudis, I., Irons, B.M., Zienkiewicz, O. (1968). Curved, isoparametric, "quadrilateral" elements for finite element analysis. *Int. J. Solids Strucrures*, 4, 31–42.

Fourier, J. (1822). *Théorie analytique de la chaleur*. Firmin Didot, Sceaux.

Fraeijs de Veubeke, B.M. and Hogge, M. (1972). Dual analysis for heat conduction problems by finite elements. *International Journal for Numerical Methods in Engineering*, 3, 65–92.

Fraeijs de Veubeke, B.M., Sander, G., Beckers, P. (1972). Dual analysis by finite elements linear and nonlinear applications [Online]. Available at: https://www.academia.edu/121169000/Dual_Analysis_by_Finite_Elements_Linear_and_Non_Linear_Applications.

Hogge, M. (1970). Analyse numérique des problèmes thermiques en construction aéronautique et spatiale. Report SF-10, Laboratoire de Techniques Aéronautiques et Spatiales, Université de Liège, Liège.

Howell, J.R., Siegel, R., Menguc, M.P. (2010). *Thermal Radiation Heat Transfer*. CRC Press, New York.

Irons, B. (1966). Numerical integration applied to finite element methods. In *Proceedings of the International Symposium on the Use of Digital Computers in Structural Engineering*. Newcastle University, Newcastle.

Lambert, J.H. (1760 [1892]). Photometrie. Photometria sive de mensura et gradibus luminis, colorum et umbrae [Online]. Available at: https://archive.org/details/bub_gb_zmpJAAAAYAAJ/page/n103/mode/2u.

Lapaine, M. (2011). Mollweide map projection. *KoG*, 15, 7–16.

Lee, H.-P. and Jackson, C.C. (1976). Finite element solution for radiative-conductive analyses with mixed diffuse specular radiation. In *Radiative Transfer and Thermal Control*, Smith, A.M. (ed.). AIAA, Reston.

Leissa, A.W. (2005). The historical bases of the Rayleigh and Ritz methods. *Journal of Sound and Vibration*, 287, 961–978.

Leopardi, P. (2006). A partition of the unit sphere into regions of equal area and small diameter. *Electron. Trans. Numer. Anal.*, 25, 309–327.

Lewis, R.W., Nithiarasu, P., Seetharamu, K.N. (2004). *Fundamentals of the Finite Element Method for Heat and Fluid Flow*. Wiley, Chichester.

Liu, B.Y.H. and Jordan, R.C. (1960). The interrelationship and characteristics distribution of direct, diffuse and total solar radiation. *Solar Energy*, 4(3), 1–19.

Nusselt, W. (1928). Graphische Bestimmung des Winkelverhaltnisses bei der Wärmestrahlung. *Zeitschrift des Vereines Deutscher Ingenieure*, 72(20), 673.

Piedallu, C. and Gégout, J.-C. (2007). Multiscale computation of solar radiation for predictive vegetation modelling. *Ann. For. Sci.*, 64, 899–909.

Planck, M. (1901). On the law of distribution of energy in the normal spectrum. *Annalen der Physik*, 4, 553.

Planck, M. (1909). *Eight Lectures on Theoretical Physics. A Course of Lectures Delivered at Columbia University*. Columbia University, New York.

Prussing, J.E. and Conway, B.A. (1993). *Orbital Mechanics*. Oxford University Press, Oxford.

Quiroz, L. and Beckers, P. (1995). Non-conforming mesh gluing in the finite elements method. *International Journal for Numerical Methods in Engineering*, 38, 2165–2184.

Ritz, W. (1911). *Œuvres complètes*. Gauthier-Villars, Paris.

Schlick, C. (1994). An inexpensive BRDF model for physically-based rendering. *Computer Graphics Forum*, 13(3), 233–246.

Schlyter, P. (1994). Computing planetary positions – A tutorial with worked examples [Online]. Available at: http://stjarnhimlen.se/english.html [Accessed 15 May 2025].

Sillion, F. and Puech, C. (1994). *Radiosity and Global Illumination*. Morgan Kaufmann, San Francisco.

Snyder, J. (1987). Map projection – A working manual. U.S. Geological Survey Professional Paper 1395. Supersedes USGS Bulletin 1532, Washington, D.C.

Sutherland, I.E., Sproull, R.F., Schumacker, R.A. (1974). Characterization of ten hidden surface algorithms. *Computing Surveys*, 6, 1–55.

Vujicic, M.R., Lavery, N.P., Brown, S.G.R. (2006). Numerical sensitivity and view factor calculation using the Monte Carlo method. In *Proceedings of the Institution of Mechanical Engineers, Part C: Journal of Mechanical Engineering Science*, 220, 697.

Washizu, K. (1955). Application of the variational method to transient heat conduction problems. Report, IR-25-17, MIT Aeroelastic and Structures Research Laboratory.

Index

L, M

N, O

P, R

S, T

V, W

Other titles from

ISTE

in

Numerical Methods in Engineering

2024

PIRIOU Francis, CLÉNET Stéphane
Finite Element Method to Model Electromagnetic Systems in Low Frequency

2023

BOUCLIER Robin, PASSIEUX Jean-Charles
IGA: Non-Invasive Coupling with FEM and Regularization of Digital Image Correlation Problems
(Isogeometric Analysis Tools for Optimization Applications in Structural Mechanics Set – Volume 2)

CIARLET Patrick, LUNÉVILLE Eric
The Finite Element Method: From Theory to Practice

2022

BOUCLIER Robin, HIRSCHLER Thibault
IGA: Non-conforming Coupling and Shape Optimization of Complex Multipatch Structures
(Isogeometric Analysis Tools for Optimization Applications in Structural Mechanics Set – Volume 1)

DERVIEUX Alain, ALAUZET Frédéric, LOSEILLE Adrien, KOOBUS Bruno
Mesh Adaptation for Computational Fluid Dynamics 1: Continuous Riemannian Metrics and Feature-based Adaptation
Mesh Adaptation for Computational Fluid Dynamics 2: Unsteady and Goal-oriented Adaptation

GRANGE Stéphane, SALCIARINI Diana
Deterministic Numerical Modeling of Soil–Structure Interaction

2021

GENTIL Christian, GOUATY Gilles, SOKOLOV Dmitry
Geometric Modeling of Fractal Forms for CAD
(Geometric Modeling and Applications Set – Volume 5)

2020

GEORGE Paul Louis, ALAUZET Frédéric, LOSEILLE Adrien, MARÉCHAL Loïc
Meshing, Geometric Modeling and Numerical Simulation 3: Storage, Visualization and In Memory Strategies
(Geometric Modeling and Applications Set – Volume 4)

SIGRIST Jean-François
Numerical Simulation, An Art of Prediction 2: Examples

2019

DA Daicong
Topology Optimization Design of Heterogeneous Materials and Structures

GEORGE Paul Louis, BOROUCHAKI Houman, ALAUZET Frédéric, LAUG Patrick, LOSEILLE Adrien, MARÉCHAL Loïc
Meshing, Geometric Modeling and Numerical Simulation 2: Metrics, Meshes and Mesh Adaptation
(Geometric Modeling and Applications Set – Volume 2)

MARI Jean-Luc, HÉTROY-WHEELER Franck, SUBSOL Gérard
Geometric and Topological Mesh Feature Extraction for 3D Shape Analysis
(Geometric Modeling and Applications Set – Volume 3)

SIGRIST Jean-François
Numerical Simulation, An Art of Prediction 1: Theory

2017

BOROUCHAKI Houman, GEORGE Paul Louis
Meshing, Geometric Modeling and Numerical Simulation 1: Form Functions, Triangulations and Geometric Modeling
(Geometric Modeling and Applications Set – Volume 1)

2016

KERN Michel
Numerical Methods for Inverse Problems

ZHANG Weihong, WAN Min
Milling Simulation: Metal Milling Mechanics, Dynamics and Clamping Principles

2015

ANDRÉ Damien, CHARLES Jean-Luc, IORDANOFF Ivan
3D Discrete Element Workbench for Highly Dynamic Thermo-mechanical Analysis
(Discrete Element Model and Simulation of Continuous Materials Behavior Set – Volume 3)

JEBAHI Mohamed, ANDRÉ Damien, TERREROS Inigo, IORDANOFF Ivan
Discrete Element Method to Model 3D Continuous Materials (Discrete Element Model and Simulation of Continuous Materials Behavior Set – Volume 1)

JEBAHI Mohamed, DAU Frédéric, CHARLES Jean-Luc, IORDANOFF Ivan
Discrete-continuum Coupling Method to Simulate Highly Dynamic Multi-scale Problems: Simulation of Laser-induced Damage in Silica Glass (Discrete Element Model and Simulation of Continuous Materials Behavior Set – Volume 2)

SOUZA DE CURSI Eduardo
Variational Methods for Engineers with Matlab®

2014

BECKERS Benoit, BECKERS Pierre
Reconciliation of Geometry and Perception in Radiation Physics

BERGHEAU Jean-Michel
Thermomechanical Industrial Processes: Modeling and Numerical Simulation

BONNEAU Dominique, FATU Aurelian, SOUCHET Dominique
Hydrodynamic Bearings – Volume 1
Mixed Lubrication in Hydrodynamic Bearings – Volume 2
Thermo-hydrodynamic Lubrication in Hydrodynamic Bearings – Volume 3
Internal Combustion Engine Bearings Lubrication in Hydrodynamic Bearings – Volume 4

DESCAMPS Benoît
Computational Design of Lightweight Structures: Form Finding and Optimization

2013

YASTREBOV Vladislav A.
Numerical Methods in Contact Mechanics

2012

DHATT Gouri, LEFRANÇOIS Emmanuel, TOUZOT Gilbert
Finite Element Method

SAGUET Pierre
Numerical Analysis in Electromagnetics

SAANOUNI Khemais
Damage Mechanics in Metal Forming: Advanced Modeling and Numerical Simulation

2011

CHINESTA Francisco, CESCOTTO Serge, CUETO Elias, LORONG Philippe
Natural Element Method for the Simulation of Structures and Processes

DAVIM Paulo J.
Finite Element Method in Manufacturing Processes

POMMIER Sylvie, GRAVOUIL Anthony, MOËS Nicolas, COMBESCURE Alain
Extended Finite Element Method for Crack Propagation

2010

SOUZA DE CURSI Eduardo, SAMPAIO Rubens
Modeling and Convexity

2008

BERGHEAU Jean-Michel, FORTUNIER Roland
Finite Element Simulation of Heat Transfer

EYMARD Robert
Finite Volumes for Complex Applications V: Problems and Perspectives

FREY Pascal, GEORGE Paul Louis
Mesh Generation: Application to finite elements – 2nd edition

GAY Daniel, GAMBELIN Jacques
Modeling and Dimensioning of Structures

MEUNIER Gérard
The Finite Element Method for Electromagnetic Modeling

2005

BENKHALDOUN Fayssal, OUAZAR Driss, RAGHAY Said
Finite Volumes for Complex Applications IV: Problems and Perspectives

Printed and bound by CPI Group (UK) Ltd, Croydon, CR0 4YY

25/06/2026

14907692-0001